Norwegian Whaling in Newfoundland:

The Aquaforte Station and the Ellefsen Family, 1902-1908

Anthony B. Dickinson and Chesley W. Sanger

International Maritime Economic History Association

St. John's, Newfoundland
2001

ISBN 0-9730073-0-3

Norwegian Whaling in Newfoundland: The Aquaforte Station and the Ellefsen Family, 1902-1908

Published by
International Maritime Economic History Association
Maritime Studies Research Unit
Memorial University of Newfoundland
St. John's, NF A1C 5S7
Canada

Front Cover Photo: *Humber* Hunting off Aquaforte (c. 1902).

Back Cover Photo: An Immature Humpback over Fifty Feet in length and over Twenty Feet around the Body, with Fins Nearly Ten-Feet Long, at Aquaforte Station.

Photos courtesy of Kmdr. Chr. Christensen Whaling Museum, Ellefsen Papers.

TABLE OF CONTENTS

LIST OF TABLES

LIST OF FIGURES

LIST OF APPENDICES

ABOUT THE AUTHORS

ANTHONY B. DICKINSON is Professor of Biology of Memorial University of Newfoundland and Executive Director of the university's International Centre. His interest in whaling arises from time spent as a Sealing Inspector/Biologist with the Government of the Falkland Islands on South Georgia and through graduate study at the Scott Polar Research Institute at Cambridge University. He has published widely on the histories of whaling and sealing.

CHESLEY W. SANGER is Honorary Research Professor, Department of Geography, and a member of the Maritime Studies Research Unit at Memorial University of Newfoundland. He has published articles on Scottish whaling and Newfoundland and Labrador sealing/whaling in *Acadiensis*; the *International Journal of Maritime History*; *Newfoundland Studies*; *Northern Scotland*; *Polar Record*; *Scottish Geographical Magazine*; *The American Neptune*; *The Mariner's Mirror*; *The Newfoundland Quarterly* and *The Northern Mariner/Le Marin du nord*.

PREFACE

The economy and social fabric of Newfoundland and Labrador have been primarily dependent on the utilization of natural resources, especially codfish and seals. The industries which developed are relatively well understood in terms of their methods and impacts. One exception is local twentieth-century shore-station whaling, which began in 1898 as a result of the decline of the Norwegian (Finnmark) industry. Although the whale fishery did not have as significant an economic impact as cod and sealing, it did provide seasonal employment and thus important benefits to communities near the stations. The following detailed examination of the Norwegian-run whaling operation at Aquaforte not only offers a snapshot of resource utilization in Newfoundland and Labrador but also provides insight into the early twentieth-century expansion of this increasingly global industry.

Commercial whaling was highly cyclical, with each phase being characterized by stock discovery, exploitation, overexpansion, fierce competition, rapid stock deletion, application of new technologies and techniques causing further depletion, and the final cessation of operations as they became unprofitable. Modern whaling in Newfoundland and Labrador followed this pattern, with four clearly-defined cyclical stages being evident between 1898 and 1972, when the government of Canada placed a moratorium on the industry. The most significant period was from 1898 to a peak just seven years after the first whale had been captured off St. John's. The Aquaforte operation began as part of the rapid expansion of whaling which characterized the initial part of this first phase. As with other operations, this was detrimentally affected by the stock overexploitation which led to the station being closed in 1908.

This research into whaling from Aquaforte is part of a larger initiative which examines the totality of the twentieth-century whale fishery in Newfoundland and Labrador and those factors which caused it to be cyclical. Particular emphasis is placed on the seasonality of operations due to stock migration patterns and the economic forces deriving from market demand and prices for the products, especially whale oil.

Information on the industry is widely and sporadically distributed. It has been collected from archives in Newfoundland, Canada, the United States, Scotland, England and Norway, especially from government papers, reports, legislation and statistical summaries, company documents and data, public and private petitions, oil and whale bone manufacturers' lists, pamphlets, diaries, logbooks, letters and telegrams. Newspaper articles were the major primary source for developing a continuous and detailed industrial profile. They yielded much data on oil and bone prices, vessels, masters, companies, shareholders, ports, hunting grounds and catches – information which formed the basic foundation for delineating sequences of events and causative factors in the cyclical fluctuations of the industry. First-hand accounts, as well as the viewpoints, hopes and fears of active participants were also revealed, including those relevant to Aquaforte. The secondary sources on commercial whaling provide little information on Newfoundland and Labrador operations and have been used largely to place the local industry into its global context.

Commercial whaling in Newfoundland and Labrador was not an activity of large companies which preserved their records. Most were short-lived, opportuntistic endeavours by local and Norwegian investors seeking quick profits. Consequently, it was gratifying to discover the existence of records covering the entire operations of one factory – that at Aquaforte. This station was owned and developed by the Ellefsen family of Stokke, near Oslo, and was the only totally Norwegian-owned factory in Newfoundland and Labrador, although others were operated by Newfoundland-Norwegian partners.

The Ellefsen and Kiserud families of Stokke and Oslo, descendants of the Norwegian owner/operators of this factory, have carefully preserved the station records, along with other items of their extensive whaling heritage. As a result of their family connections with St. John's, the authors were invited to visit Norway to examine the collection. These "Ellefesen Papers" include business transactions, operational details, decision-making documents, personal letters, photographs, and detailed accounts of wages, costs, trades/skills, equipment, catches, processing, products and prices. The archive therefore provided the basis for a case study of the operations of early twentieth-century commercial whaling in Newfoundland and Labrador, something not possible from other primary and secondary sources. This book is that case study.

The "Introduction" places the Aquaforte operation in the context of Newfoundland and global whaling, including the resource on which the industry was based, the products which provided the profits, and some of the earlier whaling activities that brought about the change from traditional to modern operating methods. The second chapter, "Early Ellefsen Involvement in Whaling," outlines the Norwegian family's previous participation in Finnmark and Icelandic whaling and the decision to shift its primary operations to the developing Newfoundland and Labrador industry. Chapter 3 describes the Ellefsen attempt to establish a new whaling factory at Aquaforte, followed in chapter 4, "The Establishment of the Aquaforte Whaling Station, 1902," by a discussion of the early success and integration into the local community. The contribution of the family to the expansion of Newfoundland and Labrador whaling in general, and thus to the decline of local whale stocks, is the principal theme of chapter 5. "The Peak Whaling Season at Aquaforte: 1904" (chapter 6) examines the state of the local industry in its peak season, while those factors which contributed to the rapid decline of the trade and the decision of the Ellefsen family to withdraw from Newfoundland and Labrador after just six seasons of whaling are discussed in chapter 7, "The Decline of the Ellefsen Whaling Operation at Aquaforte: 1905-1908." A "Postscript" outlines both the Ellefsen involvement in whaling after the family returned to Norway and the further cycles of Newfoundland and Labrador whaling from 1908.

Funding for the more general study of the Newfoundland and Labrador whaling industry came primarily from the Social Sciences and Humanities Research Council of Canada, with additional support from the offices of the Vice-President Research, Dean of Arts, and the Institute of Social and Economic Research at Memorial University of Newfoundland. Specific work on the Aquaforte case study was thus also possible. Further assistance came from Memorial's student Summer Career Placement Program and the Cartography Laboratory in the Department of Geography. Anigraph Productions Ltd. of St. John's prepared maps and figures of the Aquaforte whaling station. In Norway, Børre and Egil Kiserud and their families, their mother Marie, born in Newfoundland in 1904, and Liv Ellefsen, daughter of Edgar, welcomed us warmly into their homes and helped translate much of the Aquaforte material then housed on the family farm at Skjærsnes, but now in the Kommandør Chr. Christensens Hvalfangstmuseum in Sandefjord. Bruce Melvin of St. John's, the grand-nephew of Eda Melvin, provided photographs and genealogical information. Dr. James Mead, Curator of Marine Mammals, Smithsonian Institution, Washington, DC, helped us uncover and access its photographic collections. R.E. Eshelman of Solomons, Maryland, and A. Elton-Morris provided additional photographs. Professor Lewis Fischer of the Department of History at Memorial University of Newfoundland generously read significant proportions of the manuscript and offered insightful comments. Sharon Wall, Marilyn Hicks, Peggy Parsons, Edwina Ryan and Kevin Hicks assisted with typing and duplicating. Margaret Hennessey, the managing editor of the *Research in Maritime History* series, provided invaluable expertise throughout.

We are also grateful to the librarians, archivists and staff at the Centre for Newfoundland Studies at the Queen Elizabeth II Library at Memorial University of Newfoundland; the International Whaling Commission, Cambridge, England; Kommandør Chr. Christensens Hvalfangstmuseum, Sandefjord, Norway; National Museum of Natural History, Smithsonian Institution, Washington, DC; the Norsk Sjøfartsmuseum and Norsk Teknisk Museum, both in Oslo; Provincial Archives of Newfoundland and Labrador, St. John's; and the Vesjtfold Fylkesmuseum, Tønsberg, Norway.

Portions of this document have appeared in different formats in other academic publications or have been presented at international and national symposia. These are referenced in the footnotes and bibliography. Portions will also appear, again in different format, in a comprehensive book on Newfoundland and Labrador whaling in the twentieth century. This manuscript does, however, provide a valuable commentary on the operations of one small part of a hitherto substantially neglected component of the maritime history of Newfoundland and Labrador and of the significant contribution made then and now by Norwegians to the socio-economic development of Britain's oldest colony and Canada's newest province.

CHAPTER 1
INTRODUCTION

Traditional Global Whaling

The Newfoundland and Labrador whaling industry began as a "traditional" open-boat and hand-harpoon fishery implemented by the Basques. The early sixteenth-century decline in their hunting in the Bay of Biscay resulted in a movement of personnel and operations to Labrador and in 1610 to Spitsbergen, where the London-based Muscovy Company established a whaling industry with their assistance. Northern whaling subsequently became dominated by the Dutch, including a Greenland Sea ice-edge operation which expanded to Davis Strait and Baffin Bay. This Arctic industry later became controlled by the English and then by whalers from the east coast of Scotland in the first half of the nineteenth century (see figure 1).[1]

[1]C. Markham, "On the Whale-Fishery of the Basque Provinces of Spain," *Proceedings of the Zoological Society of London* (13 December 1881), 969-976; S.F. Harmer, "History of Whaling," *Proceedings of the Linnean Society of London*, CXL (November 1927-May 1928), 51-95; W.A Douglas and J. Bilbao, *Amerikanuak: Basques in the New World* (Reno, NV, 1975), 51-57; S. Barkham, "The Identification of Labrador Ports in Spanish 16th-Century Documents," *Canadian Cartographer*, XIV, No. 1 (1977), 1-9; Barkham, "The Basques: Filling a Gap in Our History Between Jacques Cartier and Champlain," *Canadian Geographical Journal*, XCVI, No. 1 (1978), 8-19; J.A. Tuck and R. Grenier, "A 16th-Century Basque Whaling Station in Labrador," *Scientific American*, CCXLV, No. 5 (1981), 180-188; J.A. Tuck, "Unearthing Red Bay's Whaling History," *National Geographic*, CLXVIII, No. 1 (1985), 50-57; S. Purchas, *Hakluytus Posthumus, or Purchas, His Pilgrimes, Contayning a History of the World in Sea Voyages and Lande Travells by Englishmen and Others* (London, 1625; reprint, 20 vols., Glasgow, 1905-1907), XIV, 1-23; C. de Jong, *A Short History of Old Dutch Whaling* (Pretoria, 1978); M. Conway, *No Man's Land: A History of Spitsbergen from its Discovery in 1596 to the Beginning of Scientific Exploration of the Country* (Cambridge, 1906); Conway (ed.), *Early Dutch and English Voyages to Spitsbergen in the Seventeenth Century* (London,1904); W. Scoresby, Jr., *An Account of the Arctic Regions, with a History and Description of the Northern Whale-Fishery* (2 vols., Edinburgh, 1820; reprint, Newton Abbot, 1969); G. Edmundson, *Anglo-Dutch Rivalry During the First Half of the Seventeenth Century* (Oxford, 1911); and C.W. Sanger, "The Origins of the Scottish Northern Whale Fishery" (Unpublished PhD thesis, University of Dundee, 1985).

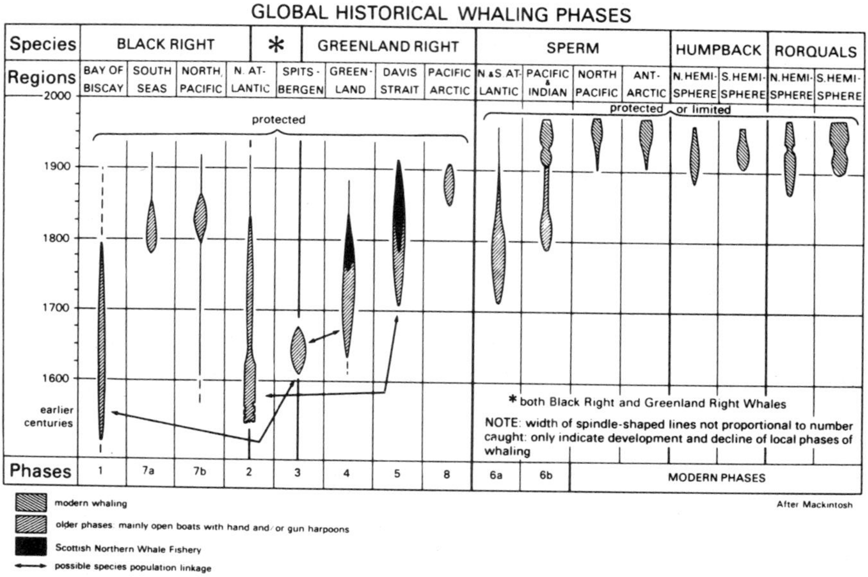

Figure 1: Historical Phases of Global Whaling

Source: Adapted by C.W. Sanger from N.A. Mackintosh, *The Stocks of Whales* (London, 1965).

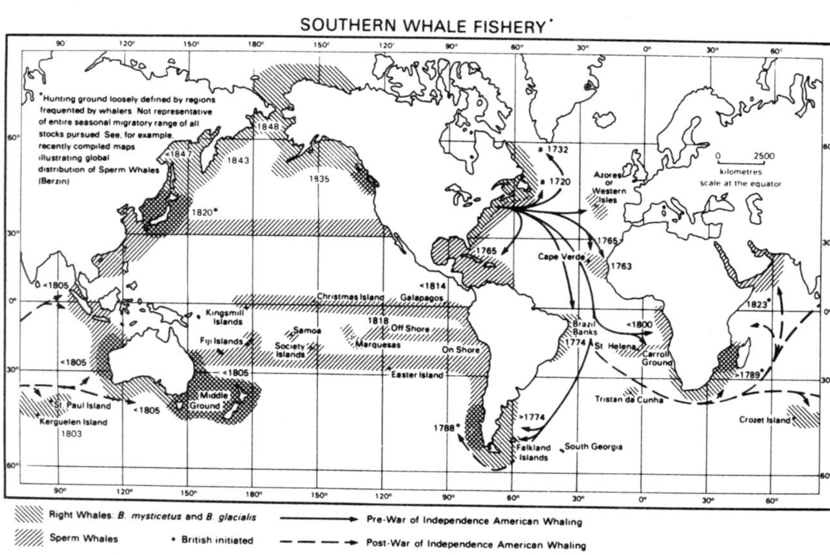

Figure 2: Southern Whale Fishery

Source: C.W. Sanger, "The Origins of the Scottish Northern Whale Fishery" (Unpublished PhD thesis, University of Dundee, 1985).

A coastal New England fishery also developed in the mid-seventeenth-century to hunt the southern right whale which, with the closely-related bowhead whale, could be harvested using the equipment and techniques developed by the Basques.[2] The colonists soon adapted these to exploit local stocks of sperm whales, resulting in a rapid expansion of the New England whaling industry during the first half of eighteenth century that subsequently involved the "discovery of new whaling grounds, the descent of a large proportion of the entire fleet upon the newly discovered regions, and their consequent exhaustion." Some of these American vessels hunted around Newfoundland from the late 1730s to about 1807.[3]

American and English whalers opened up sperm and right whale hunting grounds in the Pacific by the end of the eighteenth century, and by 1847 all the whaling grounds south of the Bering Strait had been discovered (see figure 2). New Englanders dominated global whaling by the mid-nineteenth century, so that by 1847 the estimated world whaling fleet "consisted of about 900 vessels; and of these no less than 722 belonged to the United States." This dominance was short-lived, since over-exploitation caused the entire industry to decline rapidly during the second half of the nineteenth century.[4]

---

[2]Two distinct populations of right whale (*Balaena glacialis*) exist, in temperate northern hemisphere and southern hemisphere waters. The northern populations were hunted by the Basques in the Bay of Biscay, at Spitsbergen, and off the coast of Labrador, and by the Americans along the coast of New England. Early whalers identified both as "right whales" because they were timid, slow movers that floated when killed and yielded large quantities of meat, blubber, and baleen (whalebone). The closely-related bowhead whale (*B. mysticetus*) was also taken by Europeans at Spitsbergen, in the Greenland Sea, and in Davis Strait and Baffin Bay. The taxonomic history of the genus *Balaena* and the family *Balaenidae* of the order *Mysticeti*, or whalebone whales, is discussed in L. Watson, *Sea Guide to Whales of the World* (London, 1981).

[3]C.W. Ashley, *The Yankee Whaler* (London, 1938), 41. The more comprehensive nineteenth-century accounts include C.M. Scammon, *The Marine Mammals of the North-Western Coast of North America Described and Illustrated, Together with an Account of the American Whale-fishery* (San Francisco, 1874); A. Starbuck, *History of the American Whale Fishery from its Earliest Inception to the Year 1876* (2 vols., Washington, DC, 1877); A.H. Clarke, "History and Present Condition of the Fishery," in G.B. Goode (ed.), *The Fisheries and Fishery Industries of the United States* (4 vols., Washington, DC, 1887), II, 3-218; and C.W. Sanger and A.B. Dickinson, "Whaling," in C.F. Poole (ed.), *Encyclopedia of Newfoundland and Labrador* (5 vols., St. John's, 1981-1994), V, 546-551.

[4]Following the War of Independence, the British whaler *Emelia* although "manned by the Nantucket colony of whalemen; and sailing for that ocean from London in 1787," made the first known kill of a sperm whale in the Pacific Ocean. Starbuck,

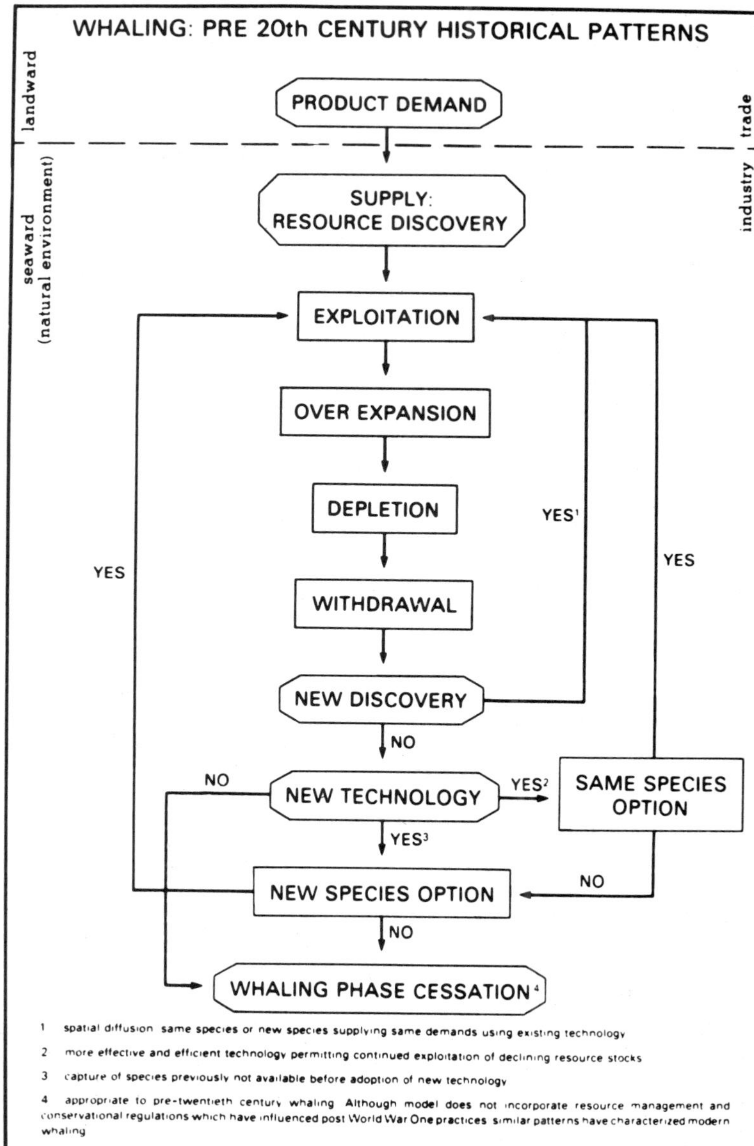

Figure 3: Whaling: Pre-Twentieth Century Historical Phases

Source: Sanger, "Origins."

History, 90; and E.P. Hohman, *The American Whaleman: A Study of Life and Labor in the Whaling Industry* (New York, 1928), 6.

The American whaling master Thomas Roys is generally credited with initiating the final phase of "pre-modern" whaling when he left Sag Harbor in July 1847 in command of the barque *Superior* for the traditional hunting grounds of the Pacific. But he was also determined to sail north through the Bering Strait if unsuccessful. *Superior* reached its "capacity of 1,800 barrels" in only thirty-five days in these northern waters, a feat which elsewhere usually required at least two seasons to accomplish. This new Arctic whaling industry soon suffered from over-exploitation and catch reduction as had previously occurred in other traditional whale fisheries (see figure 3).[5]

The adoption of innovations such as auxiliary steam-powered vessels (1880), northern shore stations (1884), and over-wintering expeditions (from 1889) allowed the industry to revive briefly. But these new practices soon placed additional pressures on a severely depleted resource, so that by the late nineteenth century virtually all stocks of bowhead, southern right, and sperm whales had been discovered, exploited, and significantly reduced. Only American bowhead whaling in the Arctic remained of "harpoon and open boat" whaling. Other large whale species remained untouched until the development of "innovative" equipment in the latter half of the century heralded the "modern" era of whaling.[6]

Modern Global Whaling

The introduction of new equipment, techniques, and hunting strategies enabled whalers to pursue the faster humpback and rorqual whales (the blue, fin, humpback and sei). During the late nineteenth century the industry therefore entered a period of unprecedented growth and change. In particular, catching was facilitated by the development of alternatives to the hand harpoon. These became especially pronounced from the late 1850s, when Thomas Roys began to use rockets to kill whales, including some from a shore station established in Iceland during the 1860s. These new technologies were subsequently refined through Norwegian initiatives led by Svend Foyn, who between 1864 and 1868 developed

---

[5]F.P. Schmitt, C. de Jong and F.H. Winter, *Thomas Welcome Roys: America's Pioneer of Modern Whaling* (Charlottesville, 1980), 28; J.R. Bockstoce, *Steam Whaling in the Western Arctic* (New Bedford, 1977); and Sanger, "Origins," 132-146.

[6]Bockstoce, *Steam Whaling*, 21-23 and 34-48. In 1857, Hull and Peterhead entrepreneurs attempted to revitalize the declining British Arctic fisheries by introducing steam-powered screw vessels into the Greenland Sea whale and seal fisheries. R.L. Webb, "Invented Too Late: The Introduction of Steam to the Arctic Whaling Fleet," *American Neptune*, XLIV (1984), 11; and C.W. Sanger, "The 19th Century Newfoundland Seal Fishery and the Influence of Scottish Whalemen," *Polar Record*, XX (1980), 235-236. See also J. Bockstoce, "History of Commercial Whaling in Arctic Alaska," *Alaska Geographic*, V, No. 4 (1978), 3; and Sanger, "Origins," 40-146.

a system that used a cannon mounted on the bow of a steam-powered catcher to fire an explosive harpoon. The line was attached to a winch-and-rope system that was strong and flexible enough to retrieve the largest rorqual (see figure 4). Working conditions in whaling were thus again dramatically transformed, and all cetaceans could now be harvested commercially.[7]

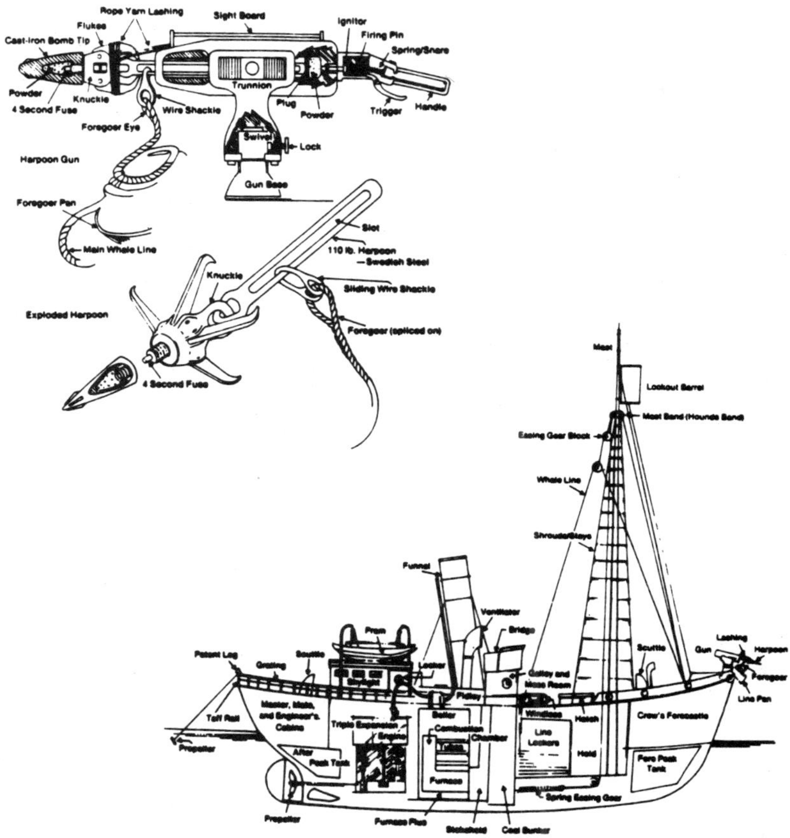

Figure 4: Modern Whaling Technology

Source: W.A. Hagelund, *Whalers No More: A History of Whaling on the West Coast* (Madeira Park, BC, 1987).

---

[7]Australia, *Whales and Whaling; Independent Inquiry Report* (2 vols., Canberra, 1979); J.N. Tønnessen and A.O. Johnsen, *The History of Modern Whaling* (Berkeley, 1982); P. Budker, *Whales and Whaling* (London, 1958), 101-121; Schmitt, de Jong and Winter, *Thomas Welcome Roys*, 63-167; Clark, *History*, 195-197; and R. Burton, *The Life and Death of Whales* (London, 1973), 132-137.

Primarily as a result of exclusive patents granted to Foyn, this original phase of "modern" whaling was first restricted to Norwegian coastal waters from a chain of shore stations along the coast of Finnmark. But it rapidly spread to new stations in the northern and southern hemispheres, quickly depleting local stocks. As a result, pelagic hunting began to re-develop in the early twentieth century using factory ships at Spitsbergen in 1904 and at South Georgia from 1907. The introduction during the 1924-1925 season of the Norwegian factory ship *Lancing*, which had a stern slipway to the main deck for onboard flensing, finally severed the industry's dependence upon shore stations and sheltered anchorages. Factory ships and their catchers could now pursue whales in any concentration or locale. This particular phase reached its zenith in the southern oceans, where exploitation of "finners" marked the real beginning of large-scale modern whaling. Shore-station whaling rapidly became unprofitable in competition with the highly-productive pelagic fleets, which brought many stocks to the verge of commercial extinction despite international regulatory efforts. By the time the International Whaling Commission (IWC) agreed in 1982 to a global moratorium on commercial whaling from 1985, most stocks had been severely depleted.[8]

Newfoundland and Labrador Whaling: Pre-1908

The Resource

Whales are totally aquatic animals, and belong to the mammalian class *cetacea*, which consists of one extinct order, *archaeoceti*, and the living orders *odontoceti* or "toothed whales" (some sixty-six species) and *mysticeti* or baleen whales (ten species). About fifteen species from both orders, including the smaller dolphins and porpoises, normally occur seasonally or year-round in the coastal waters off Newfoundland and Labrador.[9] It is difficult to determine with any precision how

---

[8]Budker, *Whales and Whaling*, 114-115; Harmer, "History," 69-70 and 72; N.A. Mackintosh, *The Stocks of Whales* (London, 1965), 145-147; D.E. Gaskin, *Whales, Dolphins and Seals, with Special Reference to the New Zealand Region* (London, 1972), 39; R.K. Headland, *The Island of South Georgia* (Cambridge, 1984), 115; R.E. Ellis, *Men and Whales* (New York, 1991), 363; Australia, *Inquiry*, 38-101; and K.R. Allen, *Conservation and Management of Whales* (London, 1980), 23-32.

[9]For discussion on the relationships of various genera and the number of species see Gaskin, *Whales, Dolphins and Seals*, 105; E. Hoyt, *The Whales of Canada* (Camden East, Ont., 1984); J. Lein, L. Fawcett, and S. Staniforth, *Wet and Fat: Whales and Seals of Newfoundland and Labrador* (St. John's, 1985); M.C. Milinkovitch, *et al.*, "Revised Phylogeny of Whales Suggested by Mitochondrial Ribosomal DNA Sequences," *Nature*, CCCLXI (1993), 346-348; M. Novacek, "Genes Tell a New Whale Tale," *Nature*, CCCLXI (1993), 298-299; K. Hay, *Guide to Identification of Whales in Newfoundland*

many were killed during the various phases of local whaling, since reasonably exact records are only available for the twentieth-century industry. These show that some 19,328 whales were taken by catchers operating from shore-stations after the beginning of local "modern" whaling. Most were fins, *Balaenoptera physalus* (12,859, or 66.5%) followed by blues, *Balaenoptera musculus* (1446, or 7.5%), humpbacks, *Megaptera novaeangliae* (1414, or 7.4%), minkes, *Balaenoptera acutorostrata* (848, or 4.4%), sperms, *Physeter catodon* (425, or 2.2%), and seis, *Balaenoptera borealis* (239, or 1.3%). Those designated as "others" (2097, or 10.9%), included 223 long-finned pilot whales, *Globicephala melaena*, six killer whales, *Orcinus orca*, one northern bottlenose whale, *Hyperoodon ampullatus*, and one southern right whale, *Eubalaena glacialis*. The balance were not recorded to the species level, especially before 1903.[10] Approximately 54,000 pilot whales were also killed during onshore small boat drives between 1948 and 1971 (see figures 5 and 6), and a Norwegian pelagic fishery took seventeen pilot, 142 minke and 824 northern bottlenose whales from the Labrador Sea between 1969 and 1972.[11]

Little objective information on the status of the commercially important cetacean species in Newfoundland and Labrador waters was available until Canada joined the International Whaling Commission (IWC) in 1949. The IWC needed information to develop catch estimates, which could only be obtained from a combination of tagging and census cruises, analyses of commercial catch reports, and fluke pattern sightings for humpback whales. These data were then used to estimate stock distribution, abundance and sustainable yield, the annual

---

Waters (St. John's, 1979); Sanger and Dickinson, "Whaling;" and L. Watson, *Sea Guide to Whales of the World* (London, 1981).

[10]*International Whaling Statistics* (Oslo, 1930-1980); and E.D. Mitchell, "Present Status of Northwest Atlantic Fin and Other Whale Stocks," in W.E. Schevill (ed.), *The Whale Problem: A Status Report* (Cambridge, MA, 1974). E.D. Mitchell and R.R. Reeves, "Catch History, Abundance, and Present Status of Northwest Atlantic Humpback Whales," in M.F. Tillman and G.P. Donovan (eds.), *Historical Whaling Records* (Cambridge, 1983), estimated 160 humpback whales processed by Newfoundland and Labrador shore stations during the period 1898-1902. These, however, are not included in this assessment of catch by species due to inaccuracy in total catch data from which the numbers were derived. Similarly, J.N. Tønnessen and A.O. Johnsen, *Den Moderne Hvalfangste Historie. Opprinelse og Utviking* (4 vols., Sandefjord, 1959-1970), observed that humpbacks comprised thirty-two percent of the 1903 catch.

[11]T. Christensen, "Preliminary Report on the Norwegian Fishery for Small Whales: Expansion of Norwegian Whaling to Arctic and Northwest Atlantic Waters and Norwegian Investigation of the Biology of Small Whales," *Journal of the Fisheries Research Board of Canada*, XXXII, No. 7 (1975), 1083-1094.

catch which can be maintained indefinitely while keeping stock size at the same level.[12] Little new information has been obtained after Canadian whaling ended, although the presumed status of several species is periodically reviewed by the Committee on the Status of Endangered Wildlife in Canada (COSEWIC). What is known about stocks of those species caught from Aquaforte (293 fin, 273 humpback, sixty-six blue and two unidentified) is summarized below, including the feeding and breeding migrations which brought them adjacent to the stations constructed in the vicinity of their routes.

Figure 5: Traditional "Pothead Drive," Chapel Arm, Trinity Bay, Newfoundland, 1955

Source: Courtesy of R.E. Eshelman, Solomons, Maryland.

Fin Whales

Fin whales were the most important catch of the modern Newfoundland and Labrador industry, including the Aquaforte station. It is the second largest of the whale species, averaging twenty-one metres in length (seventy feet) in males and

[12]For a discussion of this topic, see J.A. Gulland and L.K. Boerema, "Scientific Advice on Catch Levels," International Whaling Commission, *Twenty-third Report* (Cambridge, 1973) annex S.

twenty-two metres (seventy-three feet) in females, and is distributed worldwide. Feeding stocks concentrate in sub-Arctic and sub-Antarctic waters, then migrate south and north to equatorial seas for calving and mating in winter. They return north or south in summer to feed in the biologically-productive coastal waters of Newfoundland and Labrador. They primarily concentrate between the shore and 1000 fathoms and have a summer feeding range from 40°-57° N latitude. Newfoundland and Labrador animals feed almost exclusively on capelin, *Mallotus villosus*, and herring, *Clupea harengus harengus*.

Figure 6: Flensing: Traditional "Pothead Drive," Chapel Arm, Trinity Bay, Newfoundland, 1955

Source: See figure 5.

The Northwest Atlantic stock migrates along the Canadian continental shelf, from Cape Cod in June-July to the central Labrador coast by November. Maximum annual catches from Aquaforte occurred between June and August. Individual whales may not make the entire trip; instead latitudinally-stratified populations within the stock may move over smaller distances. Mean seasonal movement is probably not more than 600 kilometres. It has been suggested that the North Atlantic stock was between 30,000 and 50,000 before whaling began, with a Newfoundland-Labrador component of some 6000. This stock may now be around 1900. The average annual catch of 464 from 1945 to 1951 suggests that the yearly maximum sustainable yield was 400 or less, and the catch reductions which occurred periodically throughout the industry, and especially between 1951 and 1966, probably were the result of overexploitation rather than companies reducing their catching efforts as oil prices (and profits) declined. While local catch quotas were reduced from 1967 (except for the 1968 season), the quotas were never caught. Fin whales are now considered "rare" in the western North Atlantic by COSEWIC.[13]

Humpback Whales

Humpback whales occur in all oceans, the males averaging 14.6 metres (forty-eight feet) and the females 15.2 metres (fifty feet). Average weights of 30,000-40,000 kilograms (thirty-four to forty-five tons) are reported for both sexes. Mating and calving occur in warm tropical waters from 10°-20° N latitude during the winter and is followed by migration to cooler higher-latitude coastal waters for summer feeding. Feeding stocks occur in Canadian waters in the eastern North Pacific and the western North Atlantic. Humpbacks formed the second most numerous catch from Aquaforte, unlike in the larger Newfoundland and Labrador industry, where they were replaced by blue whales.

Animals in the western North Atlantic stock migrate from wintering grounds off the West Indies, particularly the Silver and Navidad Banks north of the Dominican Republic. They are seen in Newfoundland and Labrador waters especially between April and October. Most were taken from Aquaforte in May and June. Food preference varies with location and date, and although capelin are the major food off Newfoundland and Labrador, *euphausiids*, herring and squid are also consumed at different times. Changes in schooling behaviour and nutritional status of the prey markedly affect humpback distribution, and the amount of time they spend in a particular place thus also varies considerably.

Global numbers are probably between 5000 and 10,000, with the western North Atlantic stock estimated at 4000, perhaps less than eighty percent of the number before commercial whaling. Between 1535-2720 are found seasonally off Newfoundland and Labrador. IWC regulations prohibited their commercial exploitation in the North Atlantic from 1955, although forty-six were allowed to be caught for scientific purposes from 1969 to 1971, including thirty-four from

---

[13]"A Note on Baleen Whale Stocks of the Northwest Atlantic," International Whaling Commission, *Twentieth Report* (Cambridge, 1970), annex I, 1968-69; E.D. Mitchell, "Assessments of Northwest Atlantic Fin Whale Stocks," International Whaling Commission, *Twenty-second Report* (Cambridge, 1972), annex L; K.R. Allen, "Catch Per Unit Effort of Northwest Atlantic Fin Whale Stocks," International Whaling Commission, *Twenty-third Report* (Cambridge, 1973), annex H; and G.R. Meredith and R.R. Campbell, "Status of the Fin Whale, *Balaenoptera physalus*, in Canada," *Canadian Field-Naturalist*, CII, No. 2 (1988), 351-368.

stations in Newfoundland and Labrador and seven from the station at Blandford, Nova Scotia.[14]

Blue Whales

Blues are the largest whales and ranked second in numbers caught in the modern Newfoundland industry. Because of differences in migration routes, the catch from Aquaforte was significantly below that for fin and humpback whales. They have an average adult length of twenty-three metres (eighty-two feet) in males and twenty-six metres (eighty-five feet) in females and weigh between 80,000 and 130,000 kilograms (ninety to 144 tons).[15] Blue whales occur in most oceans, where they feed on swarming plankton in shallow productive higher-latitude waters and migrate to equatorial waters for breeding. Western North Atlantic stocks are found during spring, summer and fall in the Gulf of St. Lawrence, especially along the north shore from the estuary to the Strait of Belle Isle; off eastern Nova Scotia; and off southern Newfoundland. In March and April large numbers approach the Gulf of St. Lawrence from the south, keeping just outside the pack ice. They then separate as the ice clears, one group going up the estuary while the other turns east along the south coast of Newfoundland as far as Cape St. Mary's from where they move out to the Grand Banks. Most of the Aquaforte catch was taken in June. The whales do not return to the Newfoundland coast until September and October.[16]

Global numbers are probably in the vicinity of 15,000, with those in the western North Atlantic in the "very low hundreds," including sixty to one hundred in the Gulf of St. Lawrence. The western North Atlantic pre-whaling

---

[14]Michell and Reeves, "Catch History;" K.A. Hay, "Status of the Humpback Whale, *Megaptera novaeangliae*, in Canada," *Canadian Field-Naturalist*, XCIX, No. 3 (1985), 425-432; E.D. Mitchell, "Draft Report on Humpback Whales Taken under Special Scientific Permit by Eastern Canadian Land Stations, 1969-71," International Whaling Commission, *Twenty-third Report* (Cambridge, 1973), annex M; and H. Whitehead, "Updated Status of the Humpback Whale, *Megaptera novaeangliae*, in Canada," *Canadian Field-Naturalist*, CI, No. 2 (1987), 284-294.

[15]The first blue whale to be weighed was in 1903 at the Balaena whaling station in Newfoundland by Dr. F.A. Lucas, Director of the American Museum of Natural History, New York. It had a length of some seventy-seven feet and weighed sixty-three tons. See *Evening Herald* (St. John's), 11 July 1903; C.W. Sanger and A.B. Dickinson, "The Construction and Display of the First Full-Scale Model of a Blue Whale: The Newfoundland Connection," *Acadiensis*, XXVII, No. 1 (1997), 67-84; and Budker, *Whales and Whaling*, 33.

[16]J.G. Millais, *Newfoundland and its Untrodden Ways* (London, 1907), 165-166.

stock has been estimated from catch records at 1500, with a maximum annual sustainable yield of about one hundred. Commercial exploitation stopped at the end of the 1951 season in the western North Atlantic, and global protection was given in 1966. Blue whales are designated "rare" in Canadian waters by COSEWIC.[17]

Variations in seasonal inshore feeding and breeding migrations of large whale species dictated where and when they could be hunted. These patterns and stock sizes thus explain the spatial distribution and longevity of the twenty-one whaling stations, all similar in scale to Aquaforte, which periodically operated in Newfoundland and Labrador after 1898.

The Products

Pre-twentieth-century whaling was largely for oil and baleen, the filtering structures in the mouth of rorquals (or baleen whales) that is often referred to as whale bone. Prior to the development of steel and plastics, baleen, because it combined strength with elasticity, fulfilled an important if not quite so essential need as whale oil. Until the end of the nineteenth century whale bone was used primarily by the fashion industry for such items as hoop skirts, stays and hats. It was also used in a wide variety of ways: as springs for chairs, sofas and carriages; in the construction of umbrellas, whips, fishing rods, walking canes, travelling trunks and hat boxes; for making brushes; as upholstery stuffing; and in the manufacture of a broad range of minor items such as pen holders, shoe horns and combs.[18]

By the time the Aquaforte whaling station opened in 1902 whales were hunted commercially primarily for the oil which could be extracted from their blubber and other tissues. Relatively little baleen was taken. Meat was also obtained for human and animal feed, and by-products such as guano, liver and meat extract were prepared. These were of lesser importance in the modern Newfoundland and Labrador industry.

The fact that whale (and other marine) oils are unsaturated fats, and thus liquid at normal temperatures, makes them easy to transport. In the early days of whaling, oil could thus be shipped to market in wooden barrels and ultimately

---

[17]*International Whaling Statistics*, XXXII (Oslo, 1955); and A.W. Mansfield, "Status of the Blue Whale, *Balaenoptera musculus*, in Canada," *Canadian Field-Naturalist*, XCIX, No. 3 (1985), 417-420.

[18]For a full description of whale bone production and products, see Sanger, "Origins;" and Scoresby, *Account*, II.

used for lighting and heating.[19] This use declined rapidly after the introduction of coal gas and mineral oil products from the first half of the nineteenth century. In the early years of the present century whale oil was also used primarily to produce low-grade soap. A significant increase in demand occurred again after the discovery that liquid unsaturated whale oil could be converted into a saturated solid fat by hydrogenation. Perfected in 1929, this technique added hydrogen to the unsaturated fatty acids and enabled the product to be used in the production of higher-grade soap and margarine (or "butterine" as it was sometimes called).

The method of extracting oil from blubber and other body parts was essentially the same on all stations. The carcass was hauled out of the water onto a platform, and the skin and blubber were stripped off using long-handled knives and winches. The blubber was then cut into smaller pieces and hoisted up to the top of a cooker into which it was dropped and boiled. Early digestors, such as the one at Aquaforte, simply extracted oil by steam-cooking, but later modifications introduced higher pressures to make extraction more effective. After cooking, a mixture of oil, glue-water (an aqueous residue) and "grax" (solid particles) remained in the tank. When cool, the oil was skimmed off and stored, and the glue-water and grax discarded. Modification of the process later resulted in the glue-water being further treated in centrifugal separators to extract the last residues of oil. The grax and glue-water remnants were then used in the production of guano.

The fortunes of a station largely depended on its ability to produce high-quality oil and sell it when demand was high for significant profit. The oil was categorized into up to four grades based on the quantity of free fatty acids, which discoloured and added odour it. The quality depended mostly on the speed at which a carcass could be returned to the station and processed, since post-mortem decomposition by microbial activity and oxidation breaks fat down into fatty acids. Hence, the quicker the carcass could be processed, the less decomposition.

Stations strived to produce as much No. 1 oil as possible because it fetched significantly higher prices. This grade was extracted almost exclusively from blubber and had little, if any, free fatty acid content. Grades 2-4 came from blubber not rapidly processed, and from the oil yielded by the cooking of meat, bone and organs. While information is unavailable on the quality of oil produced

---

[19]Prior to the twentieth century, whale oil had a wide range of additional uses. Besides being a high-quality lubricant, it was used in the manufacture of cordage and played an important role in the preparation of rough fibres, such as jute and finer woolen clothes. It was an ingredient in paints and varnishes, and when combined with tar to form a preservative, was used extensively in the construction of wooden vessels. Whale oil also functioned as a leather softener and was used in the preparation of such widely-divergent products as perfume and explosives. Sanger, "Origins."

at Aquaforte, the highest proportion likely was No. 1 since whole carcass utilization was not practised until 1905, just three years before the station closed.

The flensed carcasses were cast out to sea, only to float ashore again to the displeasure of local residents. The recognition that a significant portion of the resource, and thus revenue, was being discarded, led to the introduction of a requirement in 1904 that such carcasses be converted into meal or guano. This was the end-product generated by the air-drying of pulverized bone, meat, intestines and other body parts through an horizontal rotating cylinder. Hot air was blown in at one end, and dried guano came out the other to be bagged. It was subsequently used as fertilizer or converted into animal feed.[20] There is evidence that the survival of shore stations in the face of increasing pelagic competition was largely due to their ability to generate additional revenues from guano production.[21]

Like other early twentieth-century whaling stations in Newfoundland and Labrador, Aquaforte was largely concerned with the production of whale oil. Its eventual demise was thus due primarily to factors that affected oil yield and sales, including a declining resource base (both in terms of total numbers and the relative abundance of higher-yielding fin and blue whales) and fluctuating demand and thus prices for oil.

Traditional Whaling

The decline of Basque shore-station whaling in Labrador was caused by the reduction of stocks due to over-hunting and migration route changes, the development of a more lucrative whale fishery at Spitsbergen, growing competition from Dutch and English whalers, and domestic strife in Spain. By the early 1580s, the overseas Basque whale fishery had virtually ended, although records exist of individual voyages being fitted out simultaneously to prosecute the Labrador whale and cod fisheries as late as the summer of 1632.[22]

Nonetheless, the western North Atlantic stock of black right whales, although reduced in size by the Basques, continued to support small-scale indigenous fisheries along the shores of New England, the Gulf of St. Lawrence

---

[20]Details on processing methods and equipment can be found in Tønnessen and Johnsen, *History*; and *Norsk Hvalfangst Tidende*, various issues.

[21]B.L. Basberg, "Survival Against all Odds? Shore Station Whaling at South Georgia in the Pelagic Era, 1925-1960," B.L. Basberg, J.E. Ringstad and E. Wexelsen (eds.), *Whaling and History: Perspectives on the Evolution of the Industry* (Sandefjord, 1993).

[22]For a fuller account of traditional whaling in Newfoundland and Labrador, see Sanger and Dickinson, "Whaling;" and Sanger, "Origins."

and perhaps even the Labrador coast to Hudson Bay and Davis Strait. French entrepreneurs also subsequently combined whaling, cod fishing and trading in the areas vacated by the Basques.

Basque whalers were replaced by New Englanders during the eighteenth century. As one American historian commented, "[i]n 1761 the fishery of the Gulf of St. Lawrence and the Straits of Belleisle was opened to our whalemen, and they speedily availed themselves of its wealth." But whales were not pursued off insular Newfoundland on any large scale until New England vessels arrived in the late 1730s. Twelve such vessels, each with a fifteen-man crew, hunted in Hermitage Bay, Bay d'Espoir and Fortune Bay between 1796 and1799 as part of an industry that continued until 1807. Thereafter American whalers only occasionally visited Newfoundland, often on their way to the more lucrative hunting grounds in Hudson Bay, Davis Strait and Baffin Bay.[23]

Although the growing eighteenth- and early nineteenth-century resident population of Newfoundland depended primarily upon the cod and seal fisheries for food and income, whale carcasses washed ashore or found floating at sea were prized by settlers for oil, baleen and meat.[24] Some of the larger nineteenth-century merchant houses in Newfoundland also undertook small-scale commercial whaling. Encouraged by the enactment of legislation in 1840 offering a £200 bounty to each of the first three vessels landing not less that ten tons of whale oil (or fifteen tons of blubber) before the tenth day of November, two St. John's companies, C.F. Bennett and Job Brothers, operated small whaling vessels along the south coast of the island between 1841 and 1844. A whaling company was also later established at Gaultois by Newman and Co. Forty to fifty whales were captured annually at the height of this operation in the late 1870s. These small domestic operations also could not be sustained, and Newman stopped whaling at the end of the century.[25]

It is therefore clear that some Newfoundlanders were engaged in seasonal inshore whaling during the eighteenth and nineteenth centuries. These relatively small-scale traditional local whaling operations, however, were soon superseded by the introduction of "modern" whaling techniques and personnel from Norway.

---

[23]Obed Macy, *The History of Nantucket: Being a Compendious Account of the First Settlement of the Island by the English, Together with the Rise and Progress of the Whale Fishery; and Other Historical Facts Relative to Said Island and Its Inhabitants* (Boston, 1835); Sanger, "Origins;" Starbuck, *History*; and Philip Tocque, *Kaleidoscope Echos: Being Historical, Philosophical, and Theological Sketches, from the Miscellaneous Writings of Philip Tocque* (Toronto, 1895).

[24]See, for example, *Evening Herald* (St. John's), 19 May and 21 August 1895; and 15 April, 9 July and 17 December 1897.

[25]*Ibid.*, 5 January 1899. See also Tocque, *Kaleidoscope Echos*.

Norwegian whaling expanded from Finnmark in a westward direction, initially to Iceland (1883) and the Faeroe Islands (1894). Newfoundland was the next target in the search for new stocks. Modern commercial whaling began in Newfoundland and Labrador following the arrival in July 1887 of Adolph Nielsen, a Norwegian government fisheries inspector who later became the first Superintendent of the new Newfoundland Department of Fisheries and a close friend of the Harvey family, important St. John's merchants. His knowledge of Newfoundland fisheries and close contacts with the Norwegian whaling community convinced the Harveys that a Newfoundland-Norwegian joint venture could make substantial profits from hunting off the coast. On 29 December 1896 they incorporated the Cabot Steam Whaling Co. Ltd. and continued hunting until 1916 to supply stations at Snook's Arm and Balaena (see figure 7). The former was the first modern shore station built in North America (see figure 8).[26]

The initial success of the Cabot Steam Whaling Co. Ltd. induced other Newfoundlanders to invest in a second operation, the Newfoundland Steam Whaling Co. Ltd., which operated from 1900 at Reuben's Cove and Rose-au-Rue. The immediate success of both resulted in the Newfoundland government receiving forty-five applications between July 1902 and December 1903 either to construct stations or renew existing licenses.[27]

The total annual catch increased rapidly (see figure 9) with the infusion of $550,000 of private capital. The industry reached its peak in 1904 when fourteen factories processed 1275 whales to produce 1,492,456 gallons of oil, 2903 tons of whale bone, and 3511 tons of guano. All eight companies were funded by local capital, except for the Cabot Steam Whaling Co. Ltd. and the Atlantic Whaling and Manufacturing Co. Ltd. The latter was capitalized entirely by Norwegian investors and operated from a station at Aquaforte on the Avalon Peninsula by the Ellefsen family of Stokke, Norway.[28]

---

[26]Tønnesen and Johnsen, *History*, 83-84; M. Baker, A.B. Dickinson and C.W. Sanger, "Adolph Nielsen: Norwegian Influence on Newfoundland Fisheries in the Late 19th-Early 20th Century," *Newfoundland Quarterly*, LXXXVII, No. 2 (1992), 25-35; and C.W. Sanger and A.B. Dickinson, "The Origins of Modern Shore-Based Whaling in Newfoundland and Labrador: The Cabot Steam Whaling Co. Ltd., 1896-1898," *International Journal of Maritime History*, I, No. 1 (1989), 129-157.

[27]A.B. Dickinson and C.W. Sanger, "Modern Shore-based Whaling in Newfoundland and Labrador: Expansion and Consolidation, 1898-1902," *International Journal of Maritime History*, II, No. 1 (1990), 83-116; and Newfoundland, Department of Fisheries, *Annual Report* (St. John's, 1903), 47-48.

[28]C.W. Sanger and A.B. Dickinson, "Expansion of Regulated Modern Shore-Station Whaling in Newfoundland and Labrador, 1902-03," *The Northern Mariner/Le Marin du Nord*, I, No. 2 (1991), 16. The 858 whales processed at seven factories in 1903

As with all the other stations, the one at Aquaforte was purposely built close to the migration routes of the large whale species. Its seasonality and longevity were thus also largely dependent on the habits of the whales and the influence of prey distribution and environmental factors on their behaviour. Other parameters, such as the initial cost of constructing new buildings, catchers and equipment, hunting success, and variations in the price of whale oil and other products, also influenced the potential for a successful whaling operation. Despite their expertise gained from whaling in Norway and Iceland, the Ellefsens had no better success than the less-experienced Newfoundland whaling entrepreneurs. Their station at Aquaforte was thus also short-lived and was only able to operate from 1902 to 1908.

---

yielded products worth $265,962, more than doubling the industry's output over the previous season. Newfoundland, Department of Fisheries, *Annual Report* (1903). In reporting on the budget introduced in the House of Assembly, the *Evening Herald* (St. John's), 12 May 1903, noted that the whaling industry had "increased in value since 1899 from $15,528 to $293,133 in 1902-3." See also A.B. Dickinson and C.W. Sanger, "Modern Shore-Station Whaling in Newfoundland and Labrador: The Peak Season, 1904," *International Journal of Maritime History*, V, No. 1 (1993), 127-154.

Figure 7: Newfoundland and Labrador Whaling Stations, 1898-1972

Source: C.W. Sanger and A.B. Dickinson, "Renewal of Newfoundland and Labrador Shore-Station Whaling, 1918-1936." *International Journal of Maritime History*, VII, No. 1 (June 1995), 84.

Figure 8: Flensing at the Snook's Arm Whaling Station, 1899

Source: Courtesy of Smithsonian Institution, National Museum of Natural History.

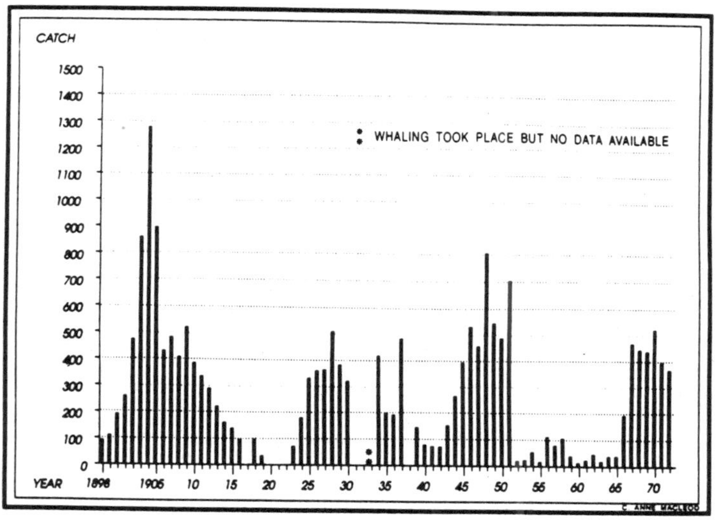

Sources: IWC, Nfld. Annual Fisheries Reports, and Newspapers (various).

Figure 9: Newfoundland Whaling Catch, 1898-1972

Source: Compiled from International Whaling Commission, *Annual Reports*, various years; Newfoundland, Department of Fisheries, *Annual Reports*; and various newspapers.

CHAPTER 2
EARLY ELLEFSEN INVOLVEMENT IN WHALING

Modern whaling started in 1868, with the refinement of Svend Foyn's technology and the resulting Norwegian dominance of the industry. A potential setback to this domination occurred, however, when the newly-established Deutsche Polar Schiffahrts Gesellschaft of Hamburg built a whaling station on the south side of Varanger Fjord opposite Foyn's factory at Vadsø (see figures 10 and 11).[1]

Foyn therefore applied directly to the King of Norway and Sweden for assistance, arguing that his catching system should be protected from foreign intervention in order to "create a new industry for Norway." His petition was successful, and on 14 January 1873 Foyn was granted exclusive whaling rights in Norwegian waters for ten years. This not only eliminated the German threat but also discouraged domestic competition. To prevent criticism that he had been given a monopoly, Foyn granted concessions to four Norwegian companies in 1880.[2] One of these, the Stokke Interessentskap for Hvalfangst, was established by Andreas Ellefsen of Stokke, a community between Tønsberg and Sandefjord on a stretch of coast which:

> provided recruits not only to practically all the crews of the Norwegian whaling fleet, but also a substantial proportion of the crews on board the whaling fleets of other nations; and for fifty years Sandefjord was the world centre for the conversion, fitting out, and equipping of whaling vessels, floating factories as well as catchers.[3]

---

[1]J.N. Tønnessen and A.O. Johnsen, *The History of Modern Whaling* (Berkeley, 1982), 30-32.

[2]*Ibid.*, 20 and 32. Foyn's role in the development of Finnmark whaling is discussed in E. Niemi, "Modern Whaling on the Norwegian Arctic Coast - Origin, Development and the Local Society," in B.L. Basberg, J.E. Ringstad and E. Wexelsen (eds.), *Whaling and History. Perspectives on the Evolution of the Industry* (Sandefjord, 1993), 67-80.

[3]Tønnessen and Johnsen, *History*, 25. The Ellefsen family, and their involvement in whaling in general, has been described by J. Falnes, *Løst Krutt og Knallpester: Hvalfangstbestyren som ble fabrikkeier i Stokke* (Sandefjord, 1999), 81-100.

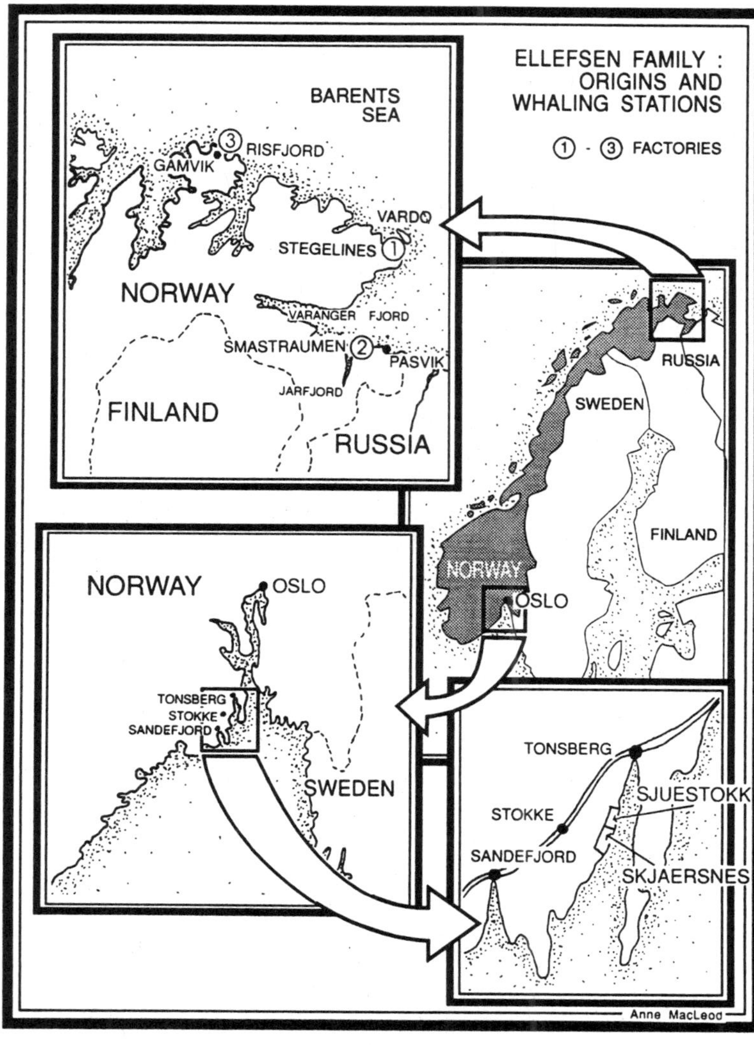

Figure 10: Ellefsen Family: Origins and Whaling Stations

Source: Courtesy of C.W. Sanger.

Figure 11: Vadsø, North Norway. Sven Foyn's whaling station is in the foreground.

Source: J.N. Tønnessen and A.O. Johnsen, *The History of Modern Whaling* (Berkeley, 1982).

Destined to become important participants in the Newfoundland whaling industry, the Ellefsen family had become affluent landowners by 1814 on the two large adjoining agricultural estates of Skjærsnes and Sjuestokk (see figures 12-15), purchased in part from the profits of maritime ventures. Some gains were also re-invested in marine activities, including the barques *Alma, Fridtjof, Oscar* and *Petrus*, and the brig *Ellef Torsen*, all built at Skjærsnes, which generated additional income and provided alternative employment for sons and cousins.[4] Andreas Ellefsen's familiarity with Foyn's commercial whaling activities, and his position as major shareholder in the Stokke Interessentskap for Hvalfangst, gave the Ellefsen family the background and incentive to become increasingly involved in the industry. The company thus spread its operations to Iceland, where its first station was established at Stegelines (1881). Others were later constructed at Småstraumen (1883-1892) and Gamvik (1892-1901) (see appendix I).

---

[4]J.K. Wogn-Henriksen, "Hvalfangst med rot i Stokke II," in *Stokke: Arv og oett* (Stokke, 1983-1984), 19.

Figure 12: Skjærsnes and Sjuestokk (upper right background).

Source: Kmdr. Chr. Christensen Whaling Museum (KCCWM), Ellefsen Papers (EP).

Figure 13: Skjærsnes, Main House (c. 1900)

Source: See figure 12.

Figure 14: Sjuestokk, Main House (c. 1900)

Source: See figure 12.

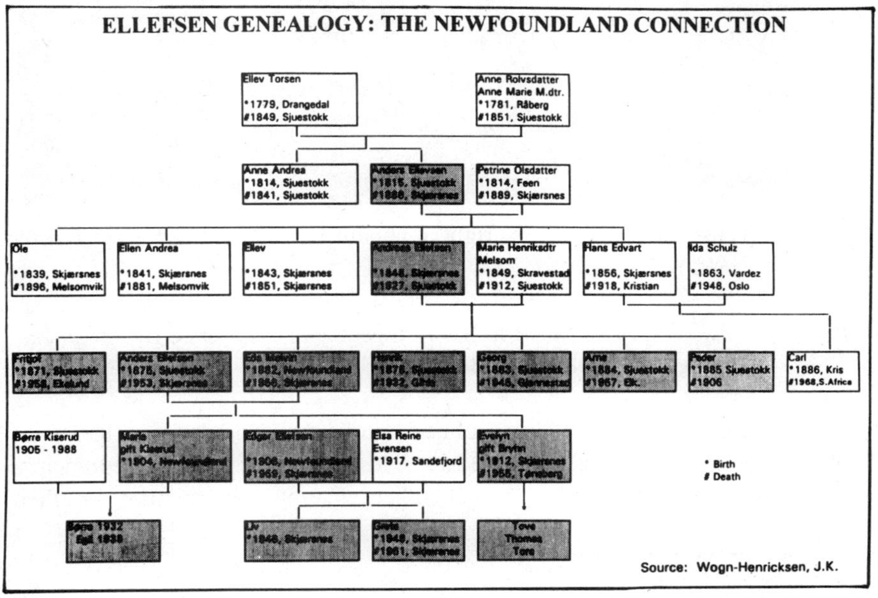

Figure 15: Ellefsen Genealogy: Newfoundland Connections

Andreas' younger brother, Hans (see figure 16), also became involved in Finnmark whaling when he acquired the Stegelines station in 1883. Although Andreas remained committed to the Finmark industry, Hans closed Stegelines in 1888 and transferred most of the equipment and the catchers *Nora* and *Ottar* to Ønundarfjødur in northwest Iceland (see appendix II). He soon dominated Icelandic whaling, including from the Mjoafjørdur station (see figure 17) after the factory at Ønundarfjødur burned down in 1901. This new station was "the largest, most efficient, and best organized in the Norwegian Sea," and at its peak processed the catch of nine vessels, yielding some 11,000 barrels of oil, more than one-quarter of all Icelandic production.[5]

Figure 16: Hans Ellefsen (extreme right) Onboard Whale Towing Vessel *Barden* (1905)

Source: J.K. Wogn-Henricksen, "Hvalfangst med rot i Stokke II," in *Stokke: Arv og oett* (Stokke, 1983-1984).

Annual Finnmark catches declined significantly from the late 1890s, and hunting was banned entirely from 1905. This led Norwegian entrepreneurs, including the Ellefsen family, to seek other whaling opportunities for their

[5]Wogn-Henriksen, "Hvalfangst," 19-29; and Tønnessen and Johnsen, *History*, 78-80.

vessels, capital and experienced crews. Hans Ellefsen sold his station and two catchers in 1911 and transferred three vessels to South Africa, where he and his son Carl established a whaling station at Saldhana Bay to harvest stocks off the Cape of Good Hope. Thirty years of Finnmark and Icelandic whaling had thus been "a nursery for the training of the crews and testing of the equipment that made it possible to develop modern catching from a local Norwegian phenomenon to a global industry."[6] The background for movement into Newfoundland had been established.

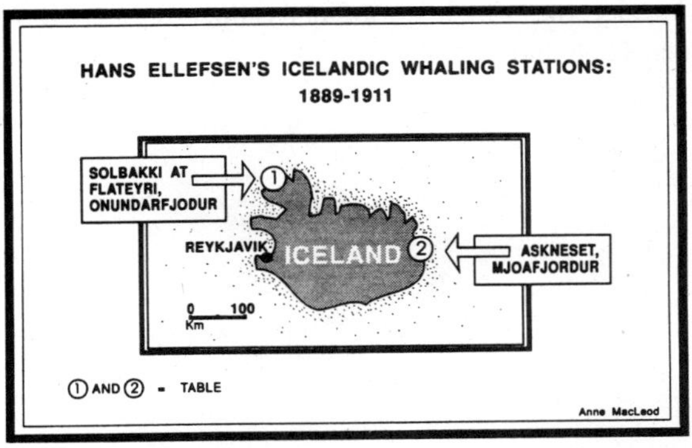

Figure 17: Ellefsen Whaling Station, Mjoafjørdur, Iceland (c. 1905)

Source: See figure 12.

---

[6]"The Norwegian government has prohibited whale fishing for a period of ten years, beginning February 1, 1905, and the sum of $100,000 has been appointed to indemnify the proprietors of whaling establishments for the loss sustained by this prohibition." *Evening Herald* (St. John's), 13 January 1905. See also Tønnessen and Johnsen, *History*, 71.

After a three-year establishment phase from 1898, the modern Newfoundland industry was now poised for rapid growth. Since Hans Ellefsen controlled Icelandic whaling at the time, and with his own Finnmark operation in decline, it is not surprising that Andreas Ellefsen was attracted by reports from Newfoundland in the summer of 1901 of large catches and rich profits.[7] Therefore, he decided to enter this rapidly- expanding North American industry as a major shareholder in the newly- constituted Atlantic Whaling and Manufacturing Co., which constructed a station at Aquaforte, the only fully Norwegian-owned of the twenty-one which periodically operated in Newfoundland and Labrador between 1898 and 1972.

---

[7]C.W. Sanger and A.B. Dickinson, "The Origins of Modern Shore-Based Whaling in Newfoundland and Labrador: The Cabot Steam Whaling Co. Ltd., 1896-1898," *International Journal of Maritime History*, I, No. 1 (1989), 129-157; Dickinson and Sanger, "Modern Shore-based Whaling in Newfoundland and Labrador: Expansion and Consolidation, 1898-1902," *International Journal of Maritime History*, II, No. 1 (1990), 83-116; Sanger and Dickinson, "Modern Shore Station Whaling in Newfoundland and Labrador: The Peak Season, 1904," *International Journal of Maritime History*, V, No. 1 (1993), 127-154; and *Evening Herald* (St. John's), 25 and 27 June; and 5, 23 and 29 July 1901.

CHAPTER 3
THE BEGINNING OF WHALING AT AQUAFORTE

The Ellefsen family was not, however, the first to catch whales from Aquaforte. Following reports of large numbers of whales to the south of St. John's, the directors of the Cabot Steam Whaling Co. Ltd. decided that *Cabot*, under the command of Captain Bull (see figure 18), should begin hunting from the sheltered harbour at Aquaforte. This decision was also influenced by the fact that their factory at Snook's Arm was still under construction. The relationship between Aquaforte and whaling therefore began when *Cabot* killed the first whale taken in the "modern" Newfoundland industry, a fin, off Baccalieu Island on 26 June 1898 and towed it back to Aquaforte for processing. These first whales killed from Aquaforte were flensed on the beach and the blubber stored on the schooner *San Juan* for transport to Snook's Arm.[1]

Figure 18: Captain Bull on *Cabot* (c. 1899)

Source: Courtesy of Smithsonian Institution, National Museum of Natural History.

[1] *Evening Herald* (St. John's), 27 June and 19 July 1898; and C.W. Sanger and A.B. Dickinson, "The Origins of Modern Shore-Based Whaling in Newfoundland and Labrador: The Cabot Steam Whaling Co. Ltd., 1896-1898," *International Journal of Maritime History*, I, No. 1 (1989), 129-157.

The experienced Norwegians on *Cabot* were able to kill whales faster than they could be flensed by the crew of *San Juan,* and another chartered schooner, *Helena May*, was sent to Aquaforte "so that the work of capture by the steamer may go on un-interruptedly." Although reports were now coming into the company "from all the northern Bays that whales [were] very plentiful," animals killed off Aquaforte contained "far more fat than those captured north." As large whales were being reported "all along the southern shore," the company decided to continue hunting from Aquaforte since the Snook's Arm station was still incomplete. Twenty-six whales were killed by *Cabot* off Aquaforte until 18 August, when they were "becoming scarce." More had by now moved into the northern bays, and *Cabot* was despatched north to the now re-opened Snook's Arm station, with *San Juan* and *Helena May* transporting the blubber from Aquaforte.[2] Aquaforte thus played a significant role in the development of this new industry, being a temporary operational base before completion of the first "modern" Newfoundland whaling station.

Figure 19: Ragnvald Berg (fourth from left, front) and Andreas Ellefsen (far right, front) on Front Steps, Sjuestokk (c. 1900)

Source: Kmdr. Chr. Christensen Whaling Museum (KCCWM), Ellefsen Papers (EP).

[2]*Evening Herald* (St. John's), 25 July; and 1, 4, 6, 18, 20 and 22 August 1898.

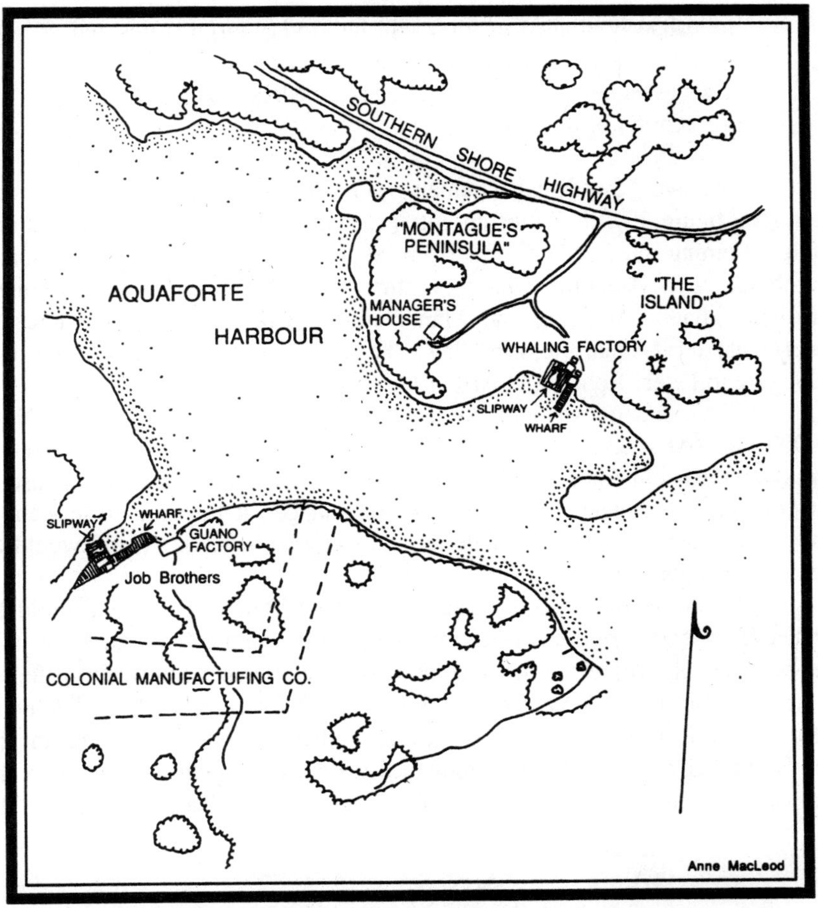

Figure 20: Location of Aquaforte Whaling Station

Source: See figure 19.

The hunting success of the Cabot Steam Whaling Co. Ltd. from Aquaforte makes it surprising that the original investors did not decide to build a station there. In fact, it was not considered seriously for the location of a station until three years later, when the St. John's *Evening Herald* reported that:

> The whale fishing industry on this coast will likely expand in the near future, and much foreign capital will be invested in it. Last week two Swedish gentlemen visited Aquaforte to negotiate for the purpose [purchase] of an extensive waterside premises there, to make a start in the enterprise as early as possible. If arrangements can be finalized for the taking over of

> the property alluded to, they will build very large stores and deep water piers, with a coal shed of a capacity for 1,000 tons. A new steamer of large size will be immediately begun for the prosecution of the voyage.[3]

The two "Swedish gentlemen" were Andreas Ellefsen and Ragnvald Berg (see figure 19), the latter an entrepreneur, owner of a large agricultural estate adjoining the Ellefsen property at Skjærsnes, and subsequently a major shareholder in the Aquaforte whaling venture. On 18 November 1901 they bought for their Atlantic Whaling and Manufacturing Co. Ltd. fourteen acres of waterfront land at Aquaforte, including "The Island" and "Montagues Peninsula," belonging to Peter, Robert and Arthur Carter (see figure 20).[4]

The company was incorporated in Norway with shares issued at NOK 15,000 each (*circa* $3750; see table 1), and the capital raised was transferred to twenty-six-year-old Anders Ellefsen, Andreas' second eldest son, for his use as construction superintendent and operations manager at the Aquaforte station. Anders was also obliged to operate the company so that this "loan" would be repaid at five percent interest.[5]

The company also contracted with Nylands Værksted of Oslo to build the catcher *Humber* (see figures 21-23) for NOK 79,000, with delivery in April 1902 (see table 2). The vessel was completed on time, and ownership was transferred to Anders Ellefsen at Aquaforte on 13 May 1902 for NOK 85,000, "delivery before arrival in Newfoundland." Unlike other catchers operating in the Newfoundland industry, *Humber* remained under Norwegian registry since the company was entirely Norwegian-owned. It was also "a shade smaller than the

---

[3]*Ibid.*, 26 October 1901.

[4]Kmdr. Chr. Chistensens Whaling Museum (KCCWM), Ellefsen Papers (EP), "Shareholder Agreement," 30 October 1901; EP, Thomas B. Clift (Notary Public and Archivist) to Andreas Ellefsen, 5 December 1906; *Evening Herald* (St. John's), 5 November 1906; *Evening Chronicle* (St. John's), 21, 25 and 28 May 1908; and Newfoundland, Registry of Companies, No. 208, Paper No. 1737, 19 January 1912.

[5]Public Archives of Newfoundland and Labrador (PANL), Job Papers, W.C. Job, "Whaling Notes Compiled by William C. Job on a Visit to Norway, June 1903," typescript; KCCWM, EP, "Shareholder Agreement," 30 October 1901; and J.K. Wogn-Henriksen, "Hvalfangst med rot i Stokke II," in *Stokke: Arv og oett* (Stokke, 1983-1984), 19 and 21. Anders had trained as a gold/silversmith and had worked in Tønsberg, Oslo, Germany, Spain and the United States prior to becoming involved in the family's Finnmark whaling operation.

other whale steamers [operating in Newfoundland], being sloop rigged and 22 tons. She is very fast and capable of doing 12 knots is smooth water."[6]

Table 1
Major Shareholders in the Atlantic Whaling and Manufacturing Company Ltd., 30 October 1901

Shareholders	Occupation	Shares	Value (Kr.)
Andreas Ellefsen	Farmer/Whaler/Shipowner	2	30,000
Hans Ellefsen	Whaler	2	30,000
Fritjof Ellefsen	Whaler	1	15,000
Ragnvald Berg	Farmer/Investor	1.8	27,500
(And "5 Others")	--	6.8	102,500
TOTALS 9	--	13.6	205,000

Source: KCCWM, EP, "Shareholder Agreement," 30 October 1901.

Table 2
Construction and Financing for *Humber*

Payment (NOK)	Activity
20,000	Contract signed
20,000	Construction started
20,000	Installation of engine and boilers
19,000	Completion date: 30 April 1902
79,000	Total cost

Source: KCCWM, EP, "Contract Agreement," 16 November 1901.

[6]KCCWM, EP, "Shareholder Agreement," 30 October 1901; EP, "Contract Agreement," 16 November 1901; and 13 May and 7 June 1902. *Humber* was Nyland hull number 132, and was ninety feet x 17'10" x 11'2" with a capacity of 280 gross tons. The price included "all equipment." See EP, letter, 25 June 1903, which lists *Humber* as the property of Anders Ellefsen.

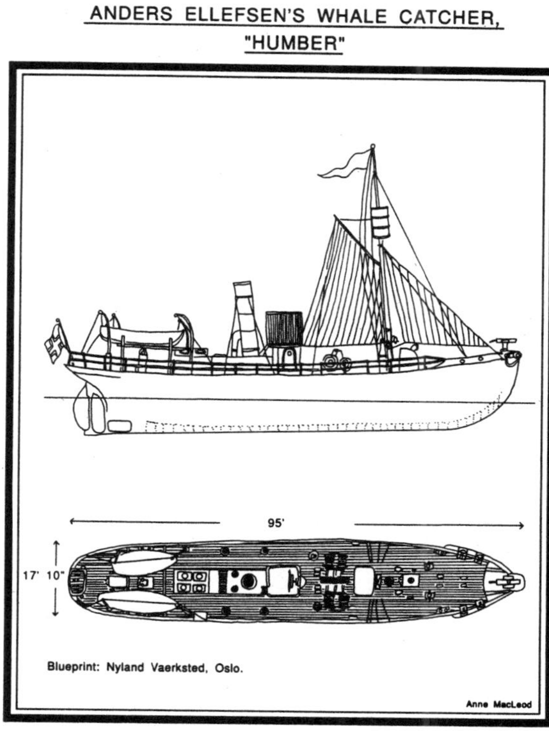

Figure 21: Blueprint of *Humber*

Source: Courtesy of Norsk Teknisk Museum, Oslo.

Figure 22: Newly-launched *Humber*, Nylands Værksted, Oslo (1902)

Source: See figure 19.

Figure 23: *Humber*, Tønsberg (1902)

Source: See figure 19.

News that a Norwegian company was becoming directly involved in the new Newfoundland whaling industry did not receive unanimous approval locally. An editorial in a St. John's newspaper, for example, argued that:

> While, theoretically, the introduction of outside capital is to be welcomed, there are instances which may prove exceptions to the rule, and in the opinion of many competent observers. *THE CASE IN POINT* appears to be one of them. We have two local companies [the Cabot Steam Whaling Co. Ltd. and the Newfoundland Steam Whaling Co. Ltd.] already engaged in this whaling industry. The capital is almost all held in this island, the profits are divided among local shareholders, and local labor is employed wherever possible. The Norwegian concern has its ships and crews ready at hand and will only invest here the cost of a factory. Whales are now fairly plentiful and this concern, with its six or eight steamers, would reap a great harvest for some years, and then could return to its own grounds when the whales here would be killed out. Our coast is limited in extent, and the regions frequented by the whales are not very great. Consequently, the competition would

> be destructive *AND WORK GREAT INJURY* to an industry which might be made of permanent benefit to the island.[7]

Similarly, a letter to the *Evening Herald* from a "Newfoundlander," complained that:

> the sooner these Norwegian gentleman understand that *the fisheries of Newfoundland belong to the fishermen of Newfoundland and not to Norway*, and that under no possible circumstance can such an enterprise be permitted here, the better for all concerned. If the Norwegians can invade this colony and catch its whales, what hinders that they should also take over the *BEST OF OUR FISHERIES*? Why should they not come and starve out our hardy toilers altogether and destroy our entire marine wealth - our cod, our herring, our lobsters, etc., "by their competition."

The writer concluded "that at all hazards the foreigners must be kept off the limited patrimony which a niggardly fate still permits us to control." Such comments were representative of the public debate "among all classes of citizens over the proposal to establish a Norwegian whaling company," with the general consensus being that "the company [should] have half its stock held here."[8]

Such sentiments obliged Anders Ellefsen to respond publicly:

> Sir, - Your editorial in this evening's issue [30 October 1901] is rather misleading, as happens to have been noticed by one who know [sic] something about it. Allow me therefore space to make a few observations on the subject. That "the foreign whaling concern on [sic] Iceland," alluded to, belongs to my uncle [Hans Ellefsen]...He has two factories, one was burned, but it is intended to be rebuilt; that he has decided to come to Nfld. is not true. I, undersigned, have come and with my own capital intend to erect a Whaling Establishment in Nfld., and subsequently to become a British subject. From parties that I have sought information, I have been well received, and even

---

[7]*Evening Herald* (St. John's), 30 October 1901.

[8]*Ibid*, 31 October and 6 November 1901.

> from high authorities have expressions of "success" been tendered to me, for which I am very thankful.[9]

Meanwhile, others were more supportive of the Ellefsen venture because it would provide new employment, increase financial benefits to local construction companies and suppliers, and generate license and tax revenues for the government. This is exemplified in a response from the Colonial Secretary to a letter from the Atlantic Whaling and Manufacturing Co. Ltd.'s local solicitor, C. O'Neill Conroy, requesting information on any requirements that would have to be satisfied before Anders Ellefsen could build and operate the Aquaforte station:

> it is perfectly competent for your client to carry on such enterprise in this colony without obtaining a license from the Government.
>
> In this Colony, a resident foreigner has as much right under the law to engage in business, as a British subject, and will receive at the hands of the Government equal protection in the conduct of his business.
>
> The Government will be very pleased indeed to encourage by every legitimate means not only your client, but any other foreigners who may be desirous of enterprising business here.[10]

The way was thus clear for the Ellefsen family and other investors in the Atlantic Whaling and Manufacturing Co. Ltd. to establish a fully Norwegian-owned and operated whaling station in Newfoundland, albeit with stringent controls on the employment of "any person not being a British subject."[11]

---

[9]*Ibid.*, 31 October 1901.

[10]KCCWM, EP, Colonial Secretary, Newfoundland, to C. O'Neill Conroy, 11 October 1901. At the time, the Newfoundland government was debating legislation to regulate the new industry. Passed on 22 April 1902, the Act required all whaling companies to pay annual license fees of $1500. Newfoundland, "An Act to Regulate the Whaling Industry," in *Acts of the General Assembly of Newfoundland, Passed in the 2nd Year of the Reign of His Majesty King Edward V* (St. John's, 1902), 101. See also C.W. Sanger and A.B. Dickinson, "Expansion of Regulated Modern Shore-Station Whaling in Newfoundland and Labrador, 1902-03," *The Northern Mariner/Le Marin du Nord*, I, No. 2 (1991).

[11]Newfoundland, "An Act to Regulate the Whaling Industry," 103.

CHAPTER 4

ESTABLISHMENT OF THE AQUAFORTE WHALING STATION, 1902

The Atlantic Whaling and Manufacturing Co. Ltd. began to construct its station in March 1902 using materials transported from St. John's to Aquaforte on the local schooners *Fashoda*, *Nonpareil* and *Mary Star*, and the coastal steamer SS *Glencoe*. The factory, coal store, oil tanks and living quarters (see figures 24-26) were projected to be completed by 25 May 1902, and as activity progressed the St. John's *Evening Herald* reported that the buildings were "large and commodious. The slip for receiving the fish [whales] is almost completed and about 50 men have been working there the past few months. Thirty permanent hands will shortly be shipped and the new steam whaler will be here by May 20th."[1]

Figure 24: Aquaforte Whaling Station (c. 1904)

Source: Kmdr. Chr. Christensen Whaling Museum (KCCWM), Ellefsen Papers (EP).

---

[1]*Evening Herald*, 8 and 12 February; 21, 22 and 24 March; 16, 19 and 30 April; and 22 and 31 May 1902. Wooden oil tanks were erected by the Horwood Lumber Co. of St. John's; coal and other supplies arrived on the barque *Helen Isabel* from New Campbellton, NB.

Figure 25: Aquaforte Whaling Station from Manager's Residence (c. 1904)

Source: See figure 24.

Figure 26: Manager's Residence, Aquaforte Whaling Station (c. 1904)

Source: KCCWM, EP, courtesy of A. Elton-Morris, St. John's.

Figure 27: Andreas Ellefsen (left) on Foredeck of *Humber*, Aquaforte Harbour (c. 1902)

Source: See figure 24.

Figure 28: *Humber* Hunting off Aquaforte (c. 1902)

Source: See figure 24.

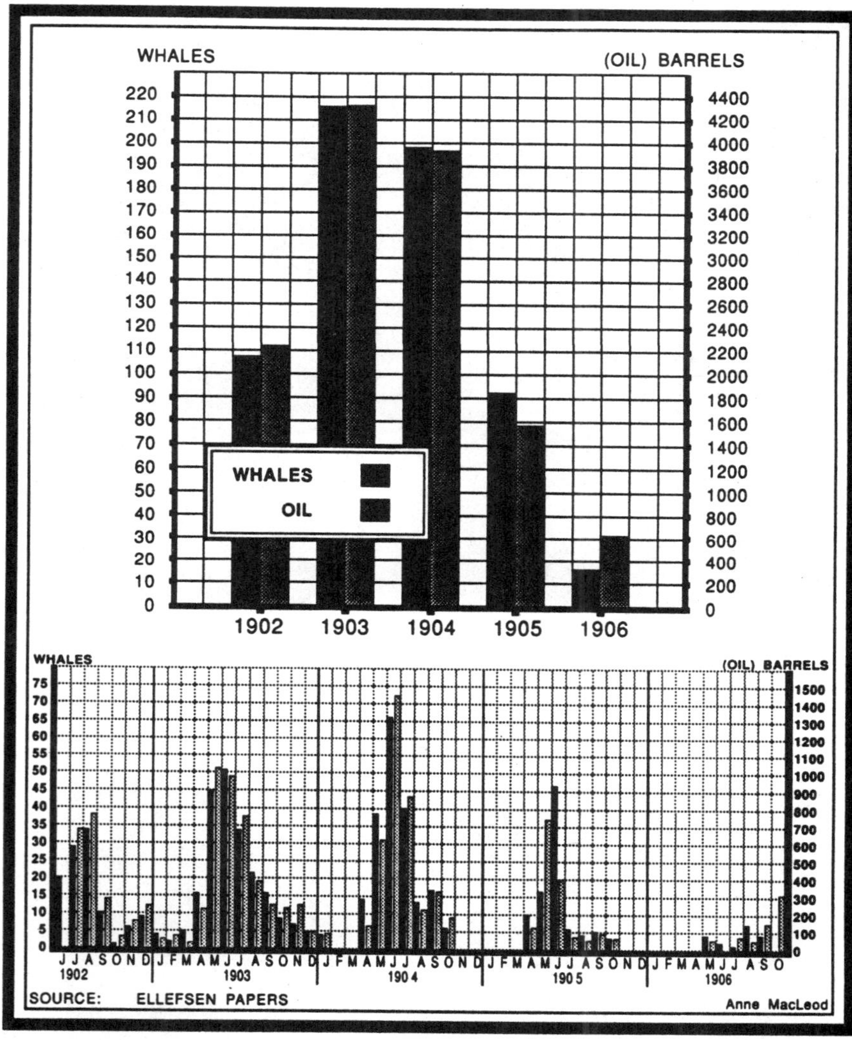

Figure 29: Catches and Oil Production, Aquaforte, 1902-1906

Source: See figure 24.

Humber arrived in St. John's from Norway on 6 June 1902 with Andreas Ellefsen and R. Lorentsen on board. The former was "on a three months visit to see after the erection of the factory and the latter is an imported cook for the crews on shore."[2] *Humber*'s first activity was transporting oil casks and other

[2]*Ibid.*, 9 June 1902.

supplies from St. John's to the still incomplete station.[3] But the presence of large numbers of whales in the immediate vicinity was too tempting for the Ellefsens, and on 17 June 1903 *Humber* killed its first whale, "with [Anders] Ellefsen's 54 year old father as gunner" (see figure 27). Two humpbacks were killed that day, which was judged to be "very good work for her first Trial and unpropitious weather, it being foggy." Ten more whales of unrecorded species were taken up to 27 June, and the carcasses moored for later processing "gave out an odour that is anything but agreeable."[4]

Ninety-three whales were killed in less than three months, mostly between Cape Race and Bay Bulls but also off Cape Bonavista (see figures 28 and 29). Local newspapers praised the new operation, and particularly related an incident in which:

> an enormous fish, apparently nothing less than 150 ft. long was harpooned. The fish rose like a small house out of the water, and rushed the steamer. It was necessary to use every precaution and manoeuvre clear of the angry monster who seemed determined to do damage. All sorts of things were tried to secure it, but, finally fearing the steamer might receive damage in one of the mad rushes, the line was cut and the ship ran away from the fish.[5]

---

[3]*Ibid.*, 11 and 14 June 1902. There seems to have been some confusion as to whether Anders Ellefsen had a local agent for *Humber*. The *Evening Herald*, 11 June 1902, recorded that A.G. Prowse held this position, whereas on 14 June it reported that "Mr. Ellefsen has appointed no agent to date and will transact the business of the company here personally."

[4]*Ibid.*, 19 and 23 April; and 18 and 27 June 1902; J.N. Tønnessen and A.O. Johnsen, *The History of Modern Whaling* (Berkeley, 1982), 104; and C.W. Sanger and A.B. Dickinson, "Expansion of Regulated Modern Shore-Station Whaling in Newfoundland and Labrador, 1902-03," *The Northern Mariner/Le Marin du Nord*, I, No. 2 (1991), 11-13. While the station operated under the provisions of the 1902 Whaling Act, it was still allowed to be built within fifty miles of another under construction at Cape Broyle by the Cape Broyle Whaling and Trading Co. Ltd.

[5]*Evening Herald* (St. John's), 4 and 11 July; and 3 and 20 September 1902. Lengths were often exaggerated; in fact, blue whales killed from Balaena averaged seventy five feet in length, See F.W. True, *The Whalebone Whales of the Western North Atlantic Compared with Those Occurring in European Waters, with Some Observations on the Species of the North Pacific* (Washington, DC, 1904), 155.

Figure 30: Balaena Whaling Station: Rissmuller's Guano Factory (right, 1901).

Source: Courtesy of Smithsonian Institution, National Museum of Natural History.

Figure 31: Whale on Slipway, Aquaforte (c. 1904)

Source: See figure 24.

Figure 32: Whale on Slipway, Aquaforte (c. 1904)

Source: See figure 24.

Figure 33: An Immature Humpback over Fifty Feet in Length and over Twenty Feet around the Body, with Fins Nearly Ten-Feet Long, at Aquaforte Station

Source: See figure 24.

A further example of such "sensational" reporting occurred one week later, when "the largest whale ever seen by the crew" was taken "6 miles off Aquaforte:"

> It was a great brute and when the harpoon entered it, it splurged so much that the hawser gave out, but it was found dead in the Bay next day. It measured 243 (?) feet and 300 casks of oil were extracted from it. Off Ferryland Head a few days ago two large whales were killed by the one harpoon as they ran side by side.[6]

Despite initial success, the Atlantic Whaling and Manufacturing Co. Ltd. experienced start-up problems, as had the two earlier companies.[7] The Newfoundland Whaling Act of 1902 required that guano production plants be built at each station so that the whole carcass could be utilized (see figure 30). This was not immediately done at Aquaforte, and the government received complaints from local residents about environmental pollution (see figures 31-33).

> Two whales are on the slip at Aquaforte and 46 to date have been put ashore at Ferryland. The people contend that the carcasses of whales should be towed away out to sea. Recently two drove in on the beach, tore up several traps and the people had to keep their windows down owing to the stench arising from them. The ship later towed them to sea.[8]

Nonetheless, most Newfoundlanders welcomed the Norwegian presence, especially since the new station significantly improved local employment opportunities, with "all the male residents of the place...working in the factory and only four boats are engaged in the cod fishery."[9] Incomes which often

---

[6]*Evening Herald*, 11 July 1902.

[7]Residents of Green Bay had informed the government of their fear that whale carcasses discarded by the Cabot Steam Whaling Co. Ltd. from its Snook's Arm station would interrupt traditional fisheries. See C.W. Sanger and A.B. Dickinson, "The Origins of Modern Shore-Based Whaling in Newfoundland and Labrador: The Cabot Steam Whaling Co. Ltd., 1896-1898," *International Journal of Maritime History*, I, No. 1 (1989), 152.

[8]*Evening Herald* (St. John's), 2 August 1902.

[9]*Ibid.*, 3 September 1902.

exceeded $20 a month (appendix III) were a welcome alternative to the uncertainties of the inshore fisheries.

Figure 34: Crew of *Humber*, and Men Employed on the Whaling Station at Aquaforte (c. 1903). James Hayward Melvin, Brother of Eda (front, 4th from right); Henrik Ellefsen (rear, 2nd from left) and Anders Ellefsen (rear right).

Source: See figure 24.

Those men from Aquaforte and neighbouring communities who worked for the company (appendices III and IV) were employed in the less skilled jobs associated with flensing and processing (see figures 35-37), and earned significantly less than the Norwegians on *Humber* (see figure 38 and appendices III and IV). This differential was consistent throughout the industry, Capt. W.E. Davidsen of the catcher *Hump*, for example, was paid $2260 for seven months work in 1910, and the crew received $300 each. By contrast, station workers received only about $25 a month. In an attempt to redress such large discrepancies, the Newfoundland Whaling Act prohibited Norwegians from crewing catchers after a station had operated for three years; the intention was that Newfoundlanders were to be trained to replace them. But the legislation did not significantly change this division of labour and status, although one Newfoundlander briefly became master of *Hawk* in 1911.[10]

---

[10]*Evening Telegram* (St. John's), 1 December 1910; Newfoundland, "An Act to Regulate the Whaling Industry," in *Acts of the General Assembly of Newfoundland, Passed in the 2nd Year of the Reign of His Majesty King Edward V* (St. John's, 1902), 103; and

Figure 35: Newfoundland Flensing Crew on Slipway, Aquaforte Whaling Station (c. 1902). Norwegian foreman, Richard Mikkelsen, left.

Source: See figure 24.

Figure 36: Newfoundland Flensing and Factory Crews, Aquaforte Whaling Station (c. 1903)

Source: See figure 24.

Evening Herald (St. John's), 20 March 1911.

Figure 37: Norwegian Crew of *Humber* (c. 1903). Capt. Henrik Ellefsen, rear left.

Source: See figure 24.

St. John's suppliers also derived considerable benefit from the Aquaforte operation, such as when *Humber* took "a large stock of meats from W. Campbell and other supplies" to the station in early September. In addition, Newfoundland as a whole benefitted from the duties paid on whale oil and bone exports, including from "the first full cargo of whale oil ever sent from this colony...for Grangemouth, Scotland "on the Norwegian steamer *Premad* in late August.[11] The first operational season, therefore, was a definite success (see figures 38 and 39) with *Humber* averaging:

> over one a day since starting, 93 whales for less than three months fishing. Last week [mid-September] was a poor one,

[11]*Evening Herald* (St. John's), 30 August; 5 and 6 September; 20 October; 22 November; and 5 December 1902. In the second half of the nineteenth century, Scottish textile manufacturers concentrated in Dundee had converted to jute processing, a procedure requiring large quantities of whale and seal oil. See C.W. Sanger, "The 19th Century Newfoundland Seal Fishery and the Influence of Scottish Whalemen," *Polar Record*, XX (1980).

only four whales being taken and three of these were killed near Cape Bonavista and towed south to the factory. The fourth was a sulphur-bottom [blue whale], and very valuable, killed on the banks off Aquaforte.[12]

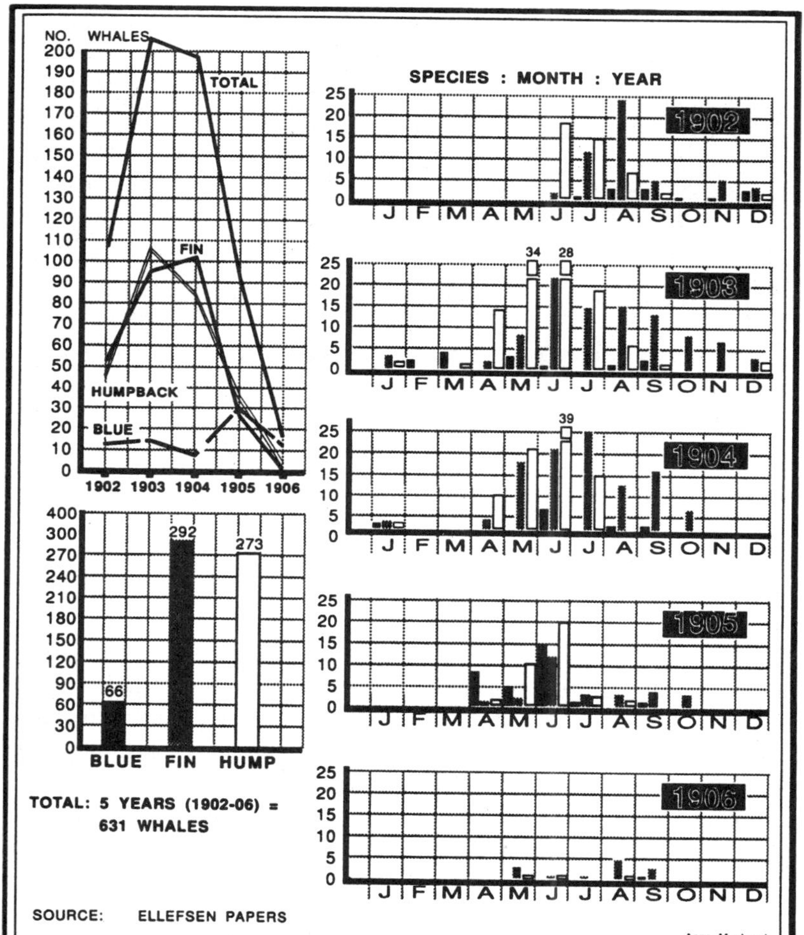

Figure 38: Species Catch, Aquaforte, 1902-1906

Source: See figure 24.

[12]*Evening Herald* (St. John's), 20 August; and 20 and 22 September 1902. A blue whale captured in late August was "valued at $700." *Humber* also hunted off the north coast of the Avalon Peninsula due to the different seasonal migration patterns of individual species. A catch of ninety-five was recorded in 1902.

OIL PER WHALE (" *Lager* " *BARRELS*)

	WHALES	OIL	AVE.
1902	108	2223	20.6
1903	216	4333	20.1
1904	198	3928	19.8
1905	92	1571	17.1
1906	17	603	35.5
	631	12,658	20.1

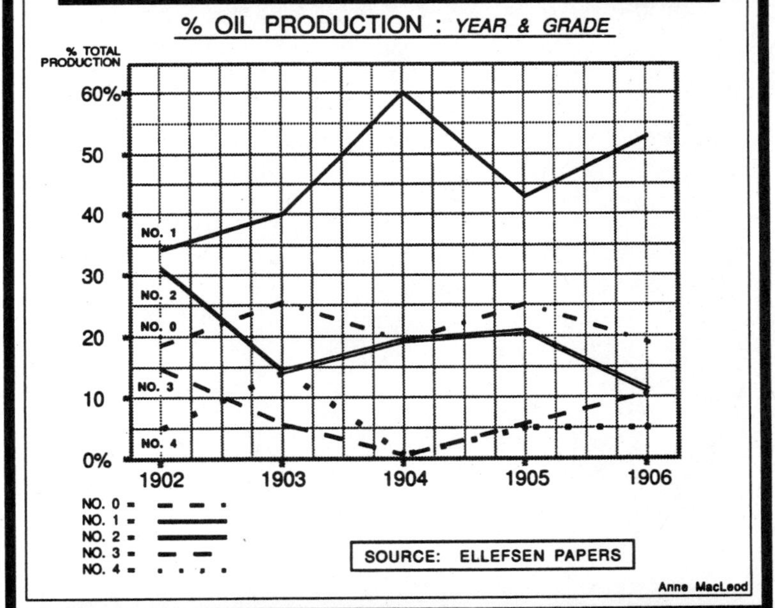

Figure 39: Annual Oil Production, Aquaforte, 1902-1906

Catches began to decline towards the end of the year, with the 108 whales killed by the end of December accounting for twenty-four percent of the 450 taken by all four vessels then hunting in a "booming" year for the expanding industry. The commitment of the Ellefsens to their Aquaforte whaling operation was further enhanced by the announcement that "[t]he wedding of Mr. Ellefsen, of the Aquaforte Whaling Co., and Miss Melvin [of St. John's], is set down for the first week in January."[13]

---

[13]*Ibid.*, 8, 9 and 27 December 1902; and Sanger and Dickinson, "Expansion," 16. Another account reported 103 killed at the beginning of the month.

CHAPTER 5
THE 1903 WHALING SEASON

Eda Hopkins Melvin was born on 23 April 1882 in St. John's, the fourth of six children of James William Melvin, an accountant with the Horwood and Co. lumber company, and Phoebe Jane (Martin) (see figures 40 and 41). James died on 7 July 1900, making it necessary for Phoebe to open her home at 8 Gower Street to boarders, including Anders Ellefsen when he visited the capital. A relationship with Eda Melvin was established when Anders hired her away from her position as a children's governess to help him learn English.[1]

The couple were married by the Rev. H.B. Cartwright at the Cathedral of St. John the Baptist in St. John's, on 5 January 1903, with Eda's brother and sister, James (see figure 42) and Beatrice Melvin, and Anders' brother Henrik among the witnesses.[2] They took up residence at Aquaforte only to suffer an immediate setback when a fire broke out on 8 January and their:

> dwelling house was completely consumed. The fire occurred through the explosion of a lamp and the occupants had barely time to save their lives. They lost everything only what they had on. Fortunately the east wind and wet weather saved the whale factory. Crowds of people gathered, and with buckets did all they could to save the house, but failed.[3]

---

[1]Bruce Melvin, St. John's, grand-nephew of Eda, personal communication, 4 June 1993.

[2]"Marriage Certificate," Possession of estate of Marie Kiserud, Oslo, Norway, daughter of Eda Hopkins Melvin and Anders Ellefsen; Kmdr. Chr. Christensens Whaling Museum (KCCWM), Ellefsen Papers (EP); and Marie Kiserud, personal communication, 1991. James Melvin subsequently worked at Aquaforte during the summer of 1903 for $10 per month. Also a witness was Eda's best friend Linda Martin, who visited her in Norway just prior to WWI. Eda did not revisit Newfoundland following her move to Norway which, according to her daughter Marie, was very difficult for her.

[3]*Evening Herald*, 9 January 1903. His St. John's lawyer wrote that he "was very sorry to hear of your fire, which occurred at a most unfortunate time. However, you were lucky to save the factory, which would have been a very serious loss." KCCWM, EP, Charles O'Neill Conroy to Anders Ellefsen, 18 February 1903.

Anders and Eda Ellefsen returned to St. John's on *Humber* to arrange for construction of a new residence, a visit not without incident, since when:

> the whaler *Humber* was berthed at the dock pier this morning one of the crew named Laurensen put down a fender to keep her from damaging the [HMS] *Calypso*. The man placed his left hand on the rail just as the vessel swung in against the warship, her full weight fell against it and the member was crushed almost to a pulp. He suffered terrible pain, and it is believed all the fingers are broken. Dr. McColloch [McGulloch] attended him shortly after noon, and it is likely he will be left here under his care.[4]

Figure 40: Eda Hopkins Melvin (c. 1900)

Source: Courtesy of B. Melvin, St. John's.

[4]*Evening Herald* (St. John's), 10 and 15 January; and 3 and 6 February 1903; and Melvin-Ellefsen Marriage Certificate. Dr. J.W. McGulloch had been a witness at Ellefsen's wedding.

Figure 41: Martin and Melvin Families (c. 1885). Front: centre, Eda Hopkins Melvin; 2nd row: left, Phoebe (Martin) Melvin; 3rd and 5th from left, Eda's grandfather and grandmother Martin; Rear: James William Melvin (left).

Source: See figure 40.

Their residence was quickly rebuilt as *Humber* continued "doing good work and Tuesday last off Cape Spear captured a fine sulphur bottom which measured 77 ft. in length. It was a female and two harpoons were discharged into its body before it was killed. It is worth about $800."[5]

Anders Ellefsen now attempted to both increase the station's profitability and eliminate complaints from fishermen and local residents about discarded carcasses, a major pollution problem during the previous season. These were now sold to Dr. Ludwig Rissmuller and towed to his new "fertilization factory," which was under construction next to the nearby whaling station at Cape Broyle (see figure 43), to produce guano from flensed carcasses of the two stations.[6]

---

[5]*Evening Herald* (St. John's), 7 February and 6 April 1903. The new house was completed on 5 April, and Eda Ellefsen returned to Aquaforte from St. John's on *Humber*.

[6]*Ibid.*, 28 February 1903. Rissmuller, a "German-American chemist," had patented a process for the whole carcass utilization of whales that was subsequently used in the Newfoundland and British Columbia industries. See C.W. Sanger and A.B.

Figure 42: James Hayward Melvin (c. 1900)

Source: See figure 40.

Although *Humber* killed a few whales during the winter and early spring, the 1903 season did not properly begin until late April. Thirty-two whales were caught by 14 May from "unusually large" numbers reported in east coast waters, and one week later *Humber* "broke all records in taking the big fish Friday. She captured 4 hump-backs and a sulphur bottom of very large size. This is the best work for one day in the history of the fishery." Good hunting continued into early

Dickinson, "Expansion of Regulated Modern Shore-Station Whaling in Newfoundland and Labrador, 1902-03," *The Northern Mariner/Le Marin du Nord*, I, No. 2 (1991); and J.M. Goddard, "The Rissmuller Factor in North American Shore Whaling, 1900-1912: The Atlantic Years," *International Journal of Maritime History*, V, No. 2 (1993), 131-155.

July, when the Aquaforte operation was recorded as "very successful." Scarcity of whales by mid-September allowed *Humber* to have its annual boiler inspection in St. John's. Then, as in the previous year, the ship resumed the hunt "in the northern bays." About one month later 200 whales had been killed, and SS *Gard* took "practically all the whale oil in the place to Glasgow, 2500 casks being stowed on board. She also took a large quantity of bone."[7]

Figure 43: Cape Broyle Whaling Station (c. 1906)

Source: Kmdr. Chr. Christensen Whaling Museum (KCCWM), Ellefsen Papers (EP).

Ellefsen returned *Humber* to the St. John's drydock in October "for repairs and general clean up and painting," as whales became even scarcer, and as poor weather and heavy sea ice later in the season made hunting difficult. At least 216 whales were processed at Aquaforte in 1903, fulfilling the promise of the previous season.[8]

---

[7]*Evening Herald* (St. John's), 6, 13, 14 and 18 May; 3 June; 2 July; 17 and 18 September; and 8 October 1903. One whale "killed at the south edge of Cape Ballard Bank, was over 90 feet long." The catch also included "one of the largest sulphur bottoms yet taken." In early June "the great fish were there [off Aquaforte] about in schools." "Bone" refers to "baleen" rather than "skeletal bone."

[8]*Ibid.*, 12 October 1903; 25 November; 18 and 22 December 1903; 9 and 28 January; and 4 and 15 February; 1904; KCCWM, EP, Anders Ellefsen, "Whale and Oil Ledger;" and The *Evening Herald*, 25 November, 1903 reported a catch of 230 whales.

Nonetheless, one problem continued to affect the operation. The Newfoundland Whaling Act of 1902 specified that the Governor-in-Council could "from time to time" enact regulations to dispose of whale carcasses and prevent them from becoming "a nuisance, or in any manner polluting the waters so as to be injurious to the public health, or to the fisheries of this Colony." By 1903 only the Aquaforte station did not have a guano-production factory attached to it as designed, constructed and operated elsewhere by Ludwig Rissmuller. Carcasses continued to be towed from Aquaforte to Rissmuller's factory at Cape Broyle. This use of the Cape Broyle factory was most likely an attempt by Ellefsen to circumvent the new regulations, as was demonstrated in his letter of 4 May 1903 to Rissmuller asking if he "would make it possible to pull four carcasses during this year from Aquaforte to Cape Broyle, so that I can tell the Government if it should enquire, that arrangements are being made to dispose of the carcasses."[9]

But heavy catching between April and August 1903 kept Ellefsen's station and catcher too busy to do this, and carcasses continued to be disposed of at sea. More complaints were made by local residents when some allegedly drifted back into Placentia Bay and fouled gear.[10] Subsequent investigations by the recently-appointed Government Whaling Inspector, Charles Way, failed to substantiate these charges, although he had visited "Mr. Ellefsen and wrote him...[that] he seemed to think the complaint was groundless."[11]

It was therefore primarily in response to the Aquaforte situation that the Deputy Colonial Secretary, Arthur Mews, sent letters to all station owners in late 1903 to inform them that:

> the Government have decided...[and]...l make it compulsory on the part of all holders of whaling licences to manufacture into

---

[9]Newfoundland, "An Act to Regulate the Whaling Industry," in *Acts of the General Assembly of Newfoundland, Passed in the 2nd Year of the Reign of His Majesty King Edward V* (St. John's, 1902), 102; and KCCWM, EP, Ellefsen to L. Rissmuller, 4 May 1903.

[10]Newfoundland, Department of Fisheries, *Annual Report* (St. John's, 1903), 4. The impact of flensed carcasses disposed of at sea from Aquaforte was provided by an American banking schooner which "ran into a carcase last month [August], at night, and the crew thought they had struck a submerged wreck until the odour assailed them. 'I can smell it yet,' remarked the captain, describing the incident a fortnight later." *Evening Herald* (St. John's), 16 September 1903.

[11]Newfoundland, Department of Fisheries, *Annual Report* (1903). Way claimed privately that "as far as I have been able to ascertain these carcasses must have certainly come from your factory at Aquaforte." KCCWM, EP, C. Way to Ellefsen, 3 September 1903.

> a commercial commodity the whole of the carcass of the whale. In other words to prevent [whaling] without having a station or factory for the conversion of the offal and carcass of the whale into oil and guano.[12]

Ellefsen's precipitating role in this is shown in a letter received from his St. John's lawyer, wherein he stated that:

> I have had information today from a private source, that the Government is contemplating closing down your factory on the ground that you have no guano factory in connection therewith. I told my informant that you had arranged with Dr. Rissmuller for the erection of a suitable plant as soon as it was possible to obtain the necessary machinery, and I hope that this will be reported in the right quarter, where it will do good.[13]

Anders Ellefsen was thus forced to introduce whole carcass utilization at Aquaforte prior to the beginning of the 1904 whaling season.

Rissmuller's agreement to purchase flensed carcasses from Ellefsen to process at Cape Broyle in part also called for him to construct a guano plant at Aquaforte, to be operational "before 1 March 1904." But the formal contract was not signed by Ellefsen until mid-October 1903, with much of the delay, and a contributory factor to his disposal of carcasses at sea during the 1903 season, resulting from his attempt to renege on the original agreement with Rissmuller.[14]

This arose from a visit to St. John's by T.R. Hunter, manager of the Huffman Manufacturing Co. of Philadelphia, a major purchaser of baleen. Hunter arrived soon after Ellefsen finalized arrangements with Rissmuller, and proposed to Ellesfen that he instead take "all of the jaw bones...at fifty dollars per ton." This better offer was too tempting for Ellefsen to refuse, despite his previous agreement. Rissmuller, in turn, took strong offense at Ellefsen's negotiations with Hunter, claiming in early June, for example, that Hunter had also approached Bowring Brothers Ltd. of St. John's, his partners in the recently-completed whaling station at Cape Broyle, "behind my back and acted in every other way in a manner no decent businessman would employ. He tried to get bone without

---

[12]KCCWM, EP, "Way to Ellefsen," 19 September 1903.

[13]*Ibid.*, Conroy to Ellefsen, 5 September 1903.

[14]*Ibid.*, Ellefsen to Rissmuller, 4 May 1903; and Conroy to Ellefsen, 10 October 1903. Rissmuller agreed to pay for carcasses at $24 for blue whales, $14 for fins and $10 for humpbacks.

me although he knew that I had stated in the States to my business friends that I had contracted for all the bones produced in Newfoundland."[15]

Aware of Rissmuller's protests, and anxious to secure a reliable supplier of bone, Hunter attempted to reassure Ellefsen that "the Dr. [Rissmuller] said he would stop you from selling me, but he can't do that until he has built his plant at your place...If you should have any trouble with the Dr. we will put up a plant at your place and use up your waste."[16] Thus reassured, Ellefsen informed Rissmuller that he considered their agreement no longer valid and that his carcasses would instead be a source of bone for Huffman. Rissmuller immediately responded that:

> I was very much surprised that you broke your agreement. I thought you might have forgotten about it, but would remember it when I reminded you about the details. I expected you would apologise, and try to make your mistake good, if it is your Power at all. Instead you dare to write me a most malicious letter full of the meanest and hideous insinuations. When you wrote it, you must not have known what you are doing...Did it ever come to your mind, that there would not be a decent price in the States for the bones in question if I would ship my output there? The demand at a high price only existed, by me keeping the masses of bones out of the market in the States. I *did not* ship a single pound there or anywhere else...I have created the demand for these bones by a large output of cash of my own and by sacrificing a good deal of my time.[17]

These "arguments" did not persuade Ellefsen to reverse his decision, and in early July 1903 he began shipping bone to Philadelphia. Unfortunately, the principals of the Huffman Co. were not prepared to honour the terms offered by Hunter, and on 23 October 1903 the company President, W.S. Huffman, informed Ellefsen that "Mr. Hunter was [only] authorized to purchase 25 tons of jaws and

---

[15]Huffman's letterhead described the company as "Manufacturers of Bone Fibre." *Ibid.*, T.R. Hunter to Ellefsen, 26 May and 2 July 1903; and Rissmuller to Ellefsen, 5 June 1903.

[16]*Ibid.*, Hunter to Ellefsen, 3 and 30 June 1903. Hunter later agreed to purchase all the bone, "if you can get Rissmuller to let you keep the jaws."

[17]*Ibid.*, Rissmuller to Ellefsen, 16 June 1903.

we [will] carry out any contract he might make to that extent."[18] Ellefsen was thus now forced to accept Rissmuller's original offer to construct a guano factory at Aquaforte.

Although he signed the contract on behalf of his Colonial Manufacturing Company soon after Ellefsen received Huffman's letter, Rissmuller did not commit to construction until February 1904. Confirming that the factory would indeed "protect the said Ellefsen from any penalty under Section 9 of the Whaling Industry Act, 1902," construction was well under way by mid-March: "Job Bros. & Co., managers of the Atlantic Mfg. Co., are erecting a new fertilizing factory in Aquaforte in connection with the whaling factory there. It was begun last week and 60 men are working on it and will hurry it to completion quickly."[19] The new factory was a substantial waterfront complex of buildings, wharfs and slipways on the south side of Aquaforte harbour opposite the whaling station (see figures 44 and 45) and was ready for the opening of the 1904 whaling season.

Figure 44: Dr. L. Rissmuller's Guano Factory, South Side of Aquaforte Harbour, Opposite Ellefsen's Whaling Station (c. 1904)

Source: See figure 43.

---

[18]*Ibid.* Hunter to Ellefsen, 16 July 1903; Ellefsen to Hunter, 24 July 1903; W.F. Huffman to Ellefsen, 29 July 1903; Huffman Manufacturing Co. to Ellefsen, 13 August and 23 October 1903.

[19]*Ibid.*, Conroy to Ellefsen, 10 October 1903; Memorial University of Newfoundland, Job Papers, No. 6, "Memorandum of Agreement," 13 February 1904; and *Evening Herald* (St. John's), 22 March 1904.

Figure 45: Aquaforte and Whaling Station from Guano Plant, foreground (c. 1904)

Source: See figure 43.

The 216 whales processed in 1903 at Aquaforte by Ellefsen's Atlantic Whaling and Manufacturing Co. Ltd. represented about one-quarter of the total Newfoundland and Labrador catch of 858 whales (see table 3). These were processed (table 4) to yield products worth at least $266,322, more than double the value of the industry from the previous year (see table 5).[20]

The early successes of the three companies now operating further convinced St. John's merchant investors that substantial profits were possible from this new industry.[21] This lead to inevitable expansion, especially encouraged by the whaling control legislation introduced in 1902.

---

[20]*Evening Herald* (St. John's), 16 March; and 19 May 1904. Forty-five tons were exported to Halifax on *Fashoda* from Snook's Arm. Opportunistic collections were also made, such as when the schooner *Shamrock* arrived at St. John's from Renews on 18 May 1904 "with 12 tons whales bones. These were taken from the coves along the shore and were part of the carcasses of fish sent adrift from the factories before the fertilizing plant was established to dispose of the bodies of the fish killed by the whalers."

[21]The Cabot Steam Whaling Co. Ltd. and the Newfoundland Steam Whaling Co. Ltd. which began operating in 1903 also did very well. Sanger and Dickinson, "Expansion."

Table 3
1903 Whaling Season Catch: Newfoundland and Labrador

		Blue	Humpback	Fin	Other	Total
Number Killed		225	287	345	1	858
Sex:	Male	-	-	-	-	485
	Female	-	-	-	-	372
	Unknown	-	-	-	-	1

Source: Newfoundland, Department of Fisheries, *Annual Report* (St. John's, 1903).

Table 4
1903 Whaling Season By-Products, Newfoundland and Labrador

Product	Tons	Value ($)
Guano	585	-
Whale Bone	773	9,950
Finners	41	-
Oil	2664	256,372
Total		**266,322**

Source: See table 3.

Table 5
Whale Oil Production and Value, and Bone Value, Newfoundland Shore-Stations, 1897/1898-1902/1903

Season	Oil			Bone	Total ($)
	Tons	Value ($)	$/ton	Value ($)	
1897-98	27	1,330	49.3	241	1,571
1898-99	211	14,439	68.4	1,167	15,606
1899-00	430	34,694	80.7	1,824	36,518
1900-01	665.25	54,221	81.5	13,550	67,771
1901-02	1,275.50	112,859	88.5	12,285	125,144
1902-03	2,664	256,372	96.3	36,761	293,133
Totals	**5,272.75**	**473,915**	**89.9**	**65,828**	**539,743**

Source: *Evening Herald* (St. John's), 14 March and 12 May 1904.

Although $550,000 of private capital was now invested in the industry, providing $19,500 for the government in licence fees during the 1902-1903 seasons, some participants began to express concern that this rate of expansion would result in the fishery being "inside of twenty years...a thing of the past." Press reports also speculated for the first time on the continued viability of the industry, since although "whaling stock is regarded as a good investment now... whether it will continue so admits of some doubt."[22]

The potential for decline was further enhanced by poor hunting practices. Crews sometimes only wounded or failed to secure their catch, and drifting carcasses were often reported.[23] Also, forty-four percent of the catch was female. The Newfoundland Department of Fisheries therefore cautioned against further industrial expansion, recommending that:

> In the true interests of this fishery, and with a view to prevent the rapid extinction of the valuable cetaceans frequenting the coasts of Newfoundland and Labrador, the Fishery Board would most strongly recommend that in construing the fifty mile clause of the Whaling Act of 1902, that a restricted constriction be put on the same. In Norway, owing to unrestricted killing, the whale fishery was depleted in fourteen years. This experience should be a guiding light to Newfoundland in the path of prudence.[24]

No attention was paid to this warning, however, and expansion of the whaling industry continued into the 1904 and 1905 seasons. Ultimately there were dire consequences, including for the facilities at Aquaforte.

---

[22]*Evening Herald* (St. John's), 30 March and 15 April 1903; and 14 March 1904. The seven factories operating in 1903 were capitalized at $350,000 and generated $10,500 in license fees for the government.

[23]*Ibid.*, 25 April; 4 June; 8 September; 12 October; and 15 December 1903.

[24]Newfoundland, Department of Fisheries, *Annual Report* (1903), 9.

CHAPTER 6
THE PEAK WHALING SEASON AT AQUAFORTE, 1904

The profits generated by the newly-established whaling companies stimulated continued expansion of the industry. The Newfoundland Steam Whaling Co. Ltd., for example, had declared a forty percent dividend and transferred a further twenty percent of profits to its reserves.[1] Recognizing the urgent need to control this expansion, the Governor of Newfoundland, Cavendish Boyle, announced in the 1904 Throne Speech to the Newfoundland House of Assembly that:

> The whale fishery has rapidly become a most important branch of our commerce, and during the present year a very large amount of additional capital will be invested therein. The value of this fishery, and the revenue that has already resulted therefrom, emphasize the wisdom of rigidly enforcing the law so wisely adopted by this Legislature two years ago for its protection.[2]

Support in the House was unanimous, since:

> This industry...under present methods, is a comparatively new one, and the returns from the money invested...have in some cases been phenomenal, while the enterprise has been the means of largely augmenting our revenues. There is every indication and every assurance at present that the coming year will see the whale fishery fleet largely increased, which will not only have the effect of giving additional employment to our people, but must necessarily swell the exchequer of our colony.[3]

---

[1]*Evening Herald* (St. John's), 26 January and 4 February 1904. Its catcher, *Puma*, killed 260 whales in 1903. C.W. Sanger and A.B. Dickinson, "Expansion of Regulated Modern Shore-Station Whaling in Newfoundland and Labrador, 1902-1903," *The Northern Mariner/Le Marin du nord*, I, No. 2 (1991), 10-11.

[2]*Evening Herald* (St. John's), 26 February 1904.

[3]*Ibid.*, 2 March 1904.

Premier Robert Bond also announced in late September that "before more licenses to capture whales are issued, enquiry should be made as to the effect of that industry upon the other fisheries of the Colony."[4] This may have been a response to ongoing but erroneous criticism from inshore fishermen that periodically reduced catches of capelin, *Mallotus villosus*, were due to "the killing out of whales," which they considered drove the fish inshore when feeding.[5] Fishermen living near the station at Rose-au-Rue, for example, reported that whereas "they used to have to steer clear of whales, now one is seldom seen."[6]

The expansion of whaling also was based on the assumptions that stocks could support an increased catching effort and that the market for oil and whalebone would continue to increase.[7] The latter was in particular demand, with 2.25 tons having been sold in Dundee for $15,000 a ton in early 1904. With only four tons reportedly left unsold globally, there was:

> no "corner" in this particular commodity. The demand has outrun the supply so rapidly that in seventy years the price of whalebone has risen from thirteen cents to nearly $7 a lb. Yet never in the greatest years of the whale fishery did any vessel make so many captures as were reported by the Newfoundland whalers in the past season. An average catch of somewhat more than two hundred whales apiece was certainly a remarkable achievement for four vessels.[8]

Many new applications for station licenses were thus received by the Newfoundland government, including an enquiry from Anders Ellefsen in late

---

[4]*Ibid.*, 28 September and 4 October 1904. Widespread scepticism existed about the ability of the inquiry to provide an objective assessment, since several government members were major shareholders in whaling companies.

[5]*Ibid.*, 12 August and 23 September 1904. Other reports were contrary; the cod fishery adjacent to stations at L'Anse au Loup and Cape Charles "has been the best known for years."

[6]*Ibid.*, 9 September 1904.

[7]*Ibid.*, 10 November 1904. The declining Davis Strait and associated fisheries provided only ten whales in 1904, and although "whalebone and oil sell high, and a small quantity makes a profitable voyage, but the boats will need to do better than they have done this year if Dundee is to keep whale fishing as one of its industries."

[8]*Ibid.*, 5 March 1904.

1903 about the possibility of establishing a second station at Hooping Harbour.[9] In the end, his St. John's lawyers recommended that he not proceed with a formal application. The government subsequently estimated optimistically that the seventeen factories projected for 1905 would yield products valued at a minimum of $700,000, more than twice that earned in 1902-1903.[10]

Table 6
Newfoundland and Labrador Whaling Production, 1904

Station	Catch	Species	Oil (gals.)	Guano (tons)	Bone (tons)
Balaena	82	57b, 9h, 11f, 5s	108,435	290	423
Snook's Arm	59	4h, 55f	57,515	138	160
Chaleur Bay	142	82b, 7h, 23f, 30s	156,100	401	324
Rose-Au-Rue	155	48b, 69f, 35h, 3s	224,560	489	387
Beaverton	28	22f, 6h	37,690	57.5	50
St. Lawrence	112	37b, 65f, 9h, 1s	192,024	502	200
Cape Broyle	150	6b, 91f, 53h	122,194	411	578
Cape Charles	104	3b, 70f, 31h	90,594	193.75	190
Dublin Cove	6	4b, 2h	6,693	4.5	5
L'Anse-Au-Loup	49	1b, 39f, 9h	53,544	51.25	90
St. Mary's	63	12b, 31f, 20h	58,590	112	100
Aquaforte	198	9b, 103f, 86h	170,821	286	160
Trinity	68	3b, 59f, 5h, 1sp	115,416	294	126
Safe Harbour	59	2b, 52f, 5h	98,280	281	110
TOTALS	**1275**	**264b, 281h, 690f, 39s,1sp**	**1,492,456**	**3511**	**2903**

Notes: b = blue, h = humpback, f = fin, s = sei, sp = sperm

Source: Newfoundland, Department of Fisheries, *Annual Report* (St. John's, 1904), 71-72.

The 1904 season started slowly because of poor weather and heavy ice.[11] Nevertheless, a record number of whales was caught, and the Aquaforte station

---

[9]*Ibid.*, 8 March 1904. The first application for a new station was from St. John's solicitor James McNeily for a for a station at Hare's Harbour, Fortune Bay.

[10]*Ibid.*, 12 May 1904; and Kmdr. Chr. Christensens Whaling Museum (KCCWM), Ellefsen Papers (EP), Charles O'Neill Conroy to Anders Ellefsen. 5 September 1903.

[11]*Evening Herald* (St. John's), 6 and 9 January; and 4 February 1904.

was the most successful (see table 6). Unfortunately, the potential for financial success was reduced by:

> a heavy glut in the whale-oil market there [Britain], no fewer than 50,000 casks of that commodity from this island being still unsold...and prospects of disposal at a good figure being quite unsatisfactory. There will not be large dividends in the whaling business this year.[12]

The market for bone still appeared to be good, however, since the Dundee Arctic whaling fleet had killed only ten and twelve whales in 1902 and 1903, respectively, producing some 16.5 tons of bone valued at £2250 and £2800 per ton.[13]

Humber captured four whales in early January, but sea ice and "strong" weather prevented continued catching. Anders Ellefsen took the opportunity to place *Humber* on drydock for general maintenance in March 1904, and towards the end of the month it was reported that the ship was again "in quest of fish."[14]

As weather conditions improved and more whales appeared on their seasonal migration, its catch increased to eighteen by the beginning of May, including nine humpbacks and fins in one week. The flensing crews "could not handle all that were taken and some were towed down to [the nearby station at] Cape Broyle." "Still doing good work," *Humber* took twenty-seven whales by mid-May, and in early June had filled Aquaforte harbour "with carcasses and the men of the factory are working overtime." By the middle of the month, *Humber*

---

[12]*Ibid.*, 7 and 13 December 1904. In addition, "the price of whale oil is now quoted at £12 stg. a ton, and no sale at that. This is a surprising drop, for last year this time it found a ready market at £24, but for some reason right this year oils have been unremarkable. It is thought by those interested that some mineral oils are taking the place of the whale product and as they can be obtained so much cheaper it has caused a slump."

[13]*Dundee Advertiser*, 3 November 1902; and 21 October; and 16 November 1903. In 1904, according to *Collier's Weekly*, "two and a quarter tons of whalebone...sold in Dundee for $15,000 a ton, and only four tons are left outstanding in the world's supply." Reported in *Evening Herald* (St. John's), 5 March 1904.

[14]*Evening Herald* (St. John's), 11 and 28 January; 15 February; and 22 and 29 March 1904. *Humber* attempted "to get through the ice" off St. John's in mid-February, but "fortunately she did not attempt going very far for the result might have been disastrous." On 7 January 1904, for example, "though the weather was very strong, she [*Humber*] was on the quest and captured a fine fish." Only four whales had been captured by the end of the January "owing to severe weather...the past fortnight."

was "high liner with 96 fish to date, for last week she secured 18, and on Saturday [18 June] alone took five."[15]

Aquaforte processed 173 whales by late August when *Humber* returned to drydock. Capturing 183 by mid-September, the ship "had every prospect of going well over the 200 mark before the work finishes," although since whales were now becoming scarce it had to "travel north of Baccalieu in quest of them."[16]

Figure 46: Eda (Melvin) Ellefsen and Daughter, Marie, Aquaforte (1904)

Source: Kmdr. Chr. Christensen Whaling Museum (KCCWM), Ellefsen Papers (EP).

---

[15]*Ibid.*, 30 April; 2, 5, and 10 May; 9 and 20 June; and 1 July 1904. One blue whale yielded eighty casks of oil.

[16]*Evening Herald* (St. John's), 27 August and 13 September 1904.

Humber again was much more successful than other catchers. Six vessels had killed 520 whales in 1904 by mid-September, an average of eighty-seven per station compared to 183 from Aquaforte.[17] This superiority may have been due to the greater expertise of *Humber*'s Norwegian crew, or to factors such as a locally greater abundance of whales. Nonetheless, *Humber* had caught 198 whales by the time hunting ended in early November. Attempts to continue whaling throughout the fall and early winter months proved unprofitable, leading Anders Ellefsen to close the station and send his Norwegian crew home.[18]

Figure 47: Henrik Ellefsen (rear left) and Wife, Hett (front left), with Anders Ellefsen and Wife, Eda, and Daughter, Marie, Aquaforte (1905)

Source: See figure 46.

It had now been almost three years since Ellefsen first arrived in Newfoundland, and in view of this he decided to return to Norway for a visit

---

[17]*Ibid.*, 16 September 1904. The *Evening Herald* incorrectly claimed 420 killed, rather than the 520 captured to that date by *Lynx* (136), *St. Lawrence* (63), *Puma* (129), *Avalon* (57), *Falcon* (100) and *Fin* (35).

[18]*Ibid.*, 10 and 15 November 1904. Fertilizer was also brought to St. John's on the schooner *Clara* (Capt. R. Shannahan).

during the seasonal lull in the industry. This also provided an opportunity to introduce his wife and new daughter, Marie (see figure 46), born at Aquaforte on 11 January 1904, to his parents and family. His brother Henrik, master of *Humber,* also married a local girl, Hett Cross (figure 47), that same year.[19] Both families therefore travelled together. The Newfoundland-born granddaughter and newly-weds were the centre of many extended family gatherings, clearly an enjoyable experience for all the Ellefsens (see figures 48 and 49). While the different language, customs and traditions of their Norwegian in-laws must have been a cultural shock to young women born and raised in the relatively isolated and unsophisticated Newfoundland society, the visit gave them with an indication of what could be expected should they permanently move to Norway.

Figure 48: Andreas Ellefsen holding Newfoundland-born Granddaughter, Marie; Family Gathering, Sjuestokk (1904-1905)

Source: See figure 46.

The Newfoundland and Labrador whaling industry was now the most important source of income and employment for Andreas Ellefsen's entire family. Arne, Georg, and Peder had joined their two older brothers in Newfoundland.

---

[19]Bruce Melvin, personal communication, 4 June 1993; KCCWM, EP, R. Berg to Anders Ellefsen, 17 July 1906; and Anders Ellefsen to Andreas Ellefsen, 4 April 1905; and *Evening Herald*, 20 March 1905.

Arne was master of *Avalon*, the catcher for the whaling station at Riverhead, St. Mary's Bay, and later took command of *Humber* while Henrik was "having a spell on shore."[20] Subsequently, he was employed in various whaling enterprises worldwide, including Japan, before retiring to Tønsberg, where he died in 1957. Georg appears to have spent only the 1905 season at Aquaforte.[21]

Figure 49: Ellefsen Family Gathering, Sjuestokk (1904-1905). Front: Anders' Newfoundland-born daughter, Marie. Rear (L-R): Anders' brother Henrik; wife Eda; brother Fritjof; brother Henrik's wife, Hett, and Aunt Kristine.

Source: See figure 46.

Peder, the youngest member of the family, worked at the Cape Broyle station in 1904 and became manager of the Riverhead station at the beginning of

[20]*Evening Herald* (St. John's), 25 September 1905. KCCWM, EP, Arne Ellefsen to M.P. Cashin, [?] August 1905.Arne Ellefsen captured three fin whales in Trinity Bay, leading the *Evening Herald* to offer congratulations: "Well done Capt. Arne, may your good luck continue."

[21]J.K. Wogn-Henriksen, "Hvalfangst med rot i Stokke II," in *Stokke: Arv og oett* (Stokke, 1983-1984), 21; and KCCWM, EP, Anders Ellefsen to Andreas Ellefsen, 21 April 1905.

the 1905 season. He was later employed by Dr. Ludwig Rissmuller as a "travelling inspector" for all of his Atlantic Whaling Company's guano plants in Newfoundland, a title Rissmuller appears to have "borrowed" from the Ellefsen operation. Peder Ellefsen returned to Norway after public accusations of improper behavior from a "Miss Blanch." He then committed "suicide on June 8 at his home in Norway, by hanging himself from a tree some twenty yards from his home [Sjuestokk]. His mind had been affected for some time, and those who knew him will learn of the news with regret and will have keen sympathy for his brothers, who reside at Aquaforte."[22]

The untimely death of this twenty-one-year-old, well-liked Norwegian surprised and saddened the Newfoundland whaling fraternity, which quickly offered condolences. His death was also obviously a tremendous blow to the close-knit Ellefsen family. Anders, writing to his lawyer in St. John's two months later, confirmed that "it was indeed a great shock to us all. My mother and father bear up better than expected."[23]

Besides their involvement at Aquaforte, the Ellefsens were also influential in determining the overall development of the Newfoundland industry, including provision of services to other whaling companies. Anders Ellefsen refilled harpoon fuses at Aquaforte, and often gave expert advice to St. John's manufacturers wishing to supply the expanding trade. As an example, the Colonial Cordage Co. Ltd., Newfoundland's largest rope supplier, turned to him for help when attempting to provide high quality whaling lines, since "we have been experimenting with the tarring of hemp whale lines and foregore [the piece of line directly attached to the harpoon], and we think we have hit upon the right thing." Some of the whaling captains, however, found them "a little too hard," leading Anders to recommend that they hire a Norwegian he knew who was fully conversant with the manufacture of whale lines and was prepared to come to St. John's to help them produce "a perfect line here."[24]

---

[22]*Evening Herald* (St. John's), 29 February 1904; 4 April 1905; and 30 June 1906; and KCCWM, EP, Peder Ellefsen to Anders Ellefsen, 22 August 1905.

[23]KCCWM, EP, P. Johnsen to Anders Ellefsen, 3 July 1906; R. Berg to Anders Ellefsen, 17 July 1906; Cashin to Anders Ellefsen, 2 July 1906; and Anders Ellefsen to C. O'Neill Conroy, 2 August 1906. Johnsen speculated that "Peder must indeed have suffered terrible anguish of mind to have caused such a sad, sad fate." Ragnvald Berg offered Anders "his sincerest condolences regarding Peder's death – it came as a bolt of lightening and will long be remembered."

[24]*Ibid.*, Mic Mac Whaling Co. to Anders Ellefsen, 19 June 1905; A. McDougall to Anders Ellefsen, 12 May 1903; Colonial Cordage Co. Ltd. to Anders Ellefsen, 26 June 1903; and T.H. Monroe to Anders Ellefsen, 15 October 1903.

Anders Ellefsen was also prepared to be the local agent for Norwegian whaling firms then expanding their businesses globally. William Job of Job Brothers Ltd., a major St. John's merchant company, took advantage of Ellefsen's presence when he visited Norway in June 1903 to arrange the construction of a catcher and to purchase equipment for shore stations he was intending to erect at Dublin Cove and L'Anse-au-Loup. His itinerary was arranged by Anders and included a letter of introduction to his father, Andreas. On returning to St. John's, Job reported that "Dr. Rissmuller has just placed an order with Nyland [shipbuilding company in Oslo] for a boat duplicate of *Viking* through Ellefsen." This helped Nyland to compete successfully later with Aker Mekaniks Værksted, the oldest and largest Norwegian shipyard. As Job reported, "Aker has built most of the whalers afloat, but of late Nyland has been cutting into them and I think chiefly by Ellefsen's help secured contracts for the *Viking* and *Humber*."[25]

The Ellefsens also took advantage of their Norwegian connections to seize business opportunities beyond whaling. Alfred Northfield, heavily involved in the expanding lumber industry on Newfoundland's northeast coast, for example, asked Anders if he could locate "any Norwegian Vessel" capable of "carrying 5 million ft. of Lumber, a distance of 160 to 180 miles between the Mills and St. John's."[26]

The family was thus significantly involved in all aspects of the Newfoundland shore-based whaling industry. Not only were they station owners and operators, but they served as factory managers, vessel captains, mates, gunners and industry inspectors. The Ellefsens also used their long-standing connections in Norwegian whaling to take full advantage of opportunities available in this rapidly-expanding Newfoundland industry.

---

[25]*Ibid.*, R.B. Job to Anders Ellefsen, 27 May 1903; and Public Archives of Newfoundland and Labrador, Job Papers, W.C. Job, "Whaling Notes Compiled by William C. Job on a Visit to Norway, June 1903," typescript, 4 and 8.

[26]KCCWM, EP, A. Northfield to Anders Ellefsen, 6 December 1907.

CHAPTER 7
THE DECLINE OF AQUAFORTE, 1905-1908

The modern Newfoundland and Labrador whaling industry peaked in 1904, when ten catchers caught 1275 whales from fourteen stations, a fifty percent increase over the previous year. These yielded 1,492,456 gallons of oil, 2903 tons of whale bone and 3511 tons of guano.[1] As a result, four new stations operated in 1905 at Hawke's Bay, Lark Harbour, Hawke Harbour, and Harbour Grace.

Preparations for the 1905 season began at Aquaforte in late March, soon after Anders and Henrik Ellefsen and their wives returned from Norway.[2] *Humber* immediately began to carry supplies to the station from St. John's and took its first whale of the season, "a fine sulphur bottom," on 18 April. Coincidently, this was both Eda's birthday and the same date as the first catch of the previous season. Both events were celebrated "very pleasantly, at Capt. Ellefsen's...Among the participants were A. Ellefsen and wife, H.C. Winsor and wife, Mrs. Tucker, Miss Robertson, James Melvin and H. Tucker."[3]

Whales appeared plentiful, as indicated in a letter Anders wrote to Ragnvald Berg in Norway four days after the first had been landed at Aquaforte:

---

[1]C.W. Sanger and A.B. Dickinson, "The Origins of Modern Shore-Based Whaling in Newfoundland and Labrador: The Cabot Steam Whaling Co. Ltd., 1896-1898," *International Journal of Maritime History*, I, No. 1 (1989); A.B. Dickinson and C.W. Sanger, "Modern Shore-based Whaling in Newfoundland and Labrador: Expansion and Consolidation, 1898-1902," *International Journal of Maritime History*, II, No. 1 (1990); C.W. Sanger and A.B. Dickinson, "Expansion of Regulated Modern Shore-Station Whaling in Newfoundland and Labrador, 1902-03," *The Northern Mariner/Le Marin du Nord*, I, No. 2 (1991); A.B. Dickinson and C.W. Sanger, "Modern Shore Station Whaling in Newfoundland and Labrador: The Peak Season, 1904," *International Journal of Maritime History*, V, No. 1 (1993).

[2]They returned to St. John's on SS *Bruce* on 19 March. *Evening Herald* (St. John's), 20 March 1905. The Aquaforte station had not been completely abandoned during their absence; the wife of Norwegian blacksmith, Hans Hang, died there on 17 February. The body was "embalmed by undertaker Carnell [St. John's] and sent to Norway for interment." *Ibid.*, 18 February 1905.

[3]*Ibid.*, 24 and 27 March; and 5, 12, 14 and 19 April 1905. *Humber* also transported oil and bone procured in 1904 to St. John's, in one instance "having in tow a small schooner laden with whale products." Other supplies, such as oil casks, arrived at the station on Job's new catcher, *Fin*.

> I should imagine that the first thing to start with is how the hunt is going, we had 4 blues in 3 days, that looks nice doesn't it? Henrik reports many whales, that is the only thing that I can say about the prospects; as you know we have never seen blues before at this time of the year, maybe the blues are coming here to us now, it looked good one week last year; well they'll be welcome and we'll try to give them a quick passage through life.[4]

An event in early June provided useful insight into the efforts sometimes required to capture the larger blue whales:

> a large sulphur bottom was captured...and quite a struggle was had before the leviathan was conquered. The weather was rough and when the harpoon hit the fish "amidships" he cut up rough and did not seem to be affected by the "shot" and succeeded in badly smashing the donkey winch on deck to which the cable had been affixed. To capture the fish three men were lowered away in a boat and after nearly two hours exertion and at much risk dispatched the fish by putting three hand lances into him. The men had life-belts on them and say that the last narrowly escaped a strike of the monster's tail in its struggles.[5]

Whales were "visible everywhere," and *Humber* secured excellent catches into May. While the early rate of capture appeared to indicate that the promise of the previous two seasons would again be fulfilled, by the end of the month it had taken only twelve percent (twenty-seven) of all the whales killed by the eight catchers then operating "on the south coast."[6]

Despite this seemingly low catch, the new Rissmuller guano factory which had been built across the harbour from the station could not handle all the flensed carcasses, a situation of considerable concern to Anders Ellefsen in view of the problems he had experienced with Newfoundland Fisheries Inspectors over

---

[4]*Ibid.*, 24 April 1905; and Kmdr. Chr. Christensens Whaling Museum (KCCWM), Ellefsen Papers (EP), Anders Ellefsen to R. Berg, 21 April 1905.

[5]*Evening Herald* (St. John's), 5 June 1905.

[6]*Ibid.*, 17, 20 and 29 May 1905. The season further north was annually delayed by sea ice. Two hundred and twenty-six whales had been killed by this date: *Lynx* (46), *Mic Mac* (38), *Cabot* (38), *Falcon* (24), *St. Lawrence* (21), *Puma* (20), and *Avalon* (12).

the previous two seasons: "[i]f this is going to continue [failure to remove blubber scraps from water around factory], I am afraid some trouble will come from it...I do not want to get into future trouble."[7]

Although whales continued to be plentiful throughout the first half of June, and were "larger than any taken last year," *Humber* and other "southern whalers" at the end of the month were reported having "not a half dozen between them." By early July, Ellefsen's station had processed only forty-six whales, "about half the total they had at this time last year." In fact, *Humber* captured only one whale between 18 June and 10 July, when it was reported that whaling was "slack south." As a result, Henrik Ellefsen then took her "north to the Funks...in quest of cetaceans, having covered the coast from Cape Pine [south of St. John's] to Cape Bonavista, for a radius of 80 miles off shore and not seen a 'spout.'" Despite these efforts, there was little improvement in August.[8]

It was not until late fall that whales reappeared along the southeast coast of the island, within hunting range of Aquaforte. Their small numbers made it necessary for *Humber* to continue searching further afield, resulting in the capture of a fin whale off Cape Bonavista on 3 October. Two weeks later "a large fish" was also taken deep in Conception Bay. These longer voyages were more expensive, reduced profit margins significantly, and resulted in lower quality oil production. Only sixty-five whales, barely thirty percent of the 1904 catch, had been killed from Aquaforte when the 1905 season ended in early October.[9]

The migratory stocks apparently could not support the high catches of the previous seasons and the increased level of exploitation brought on by the opening of new stations. The 1905 catch declined by more than thirty percent to 892 (263 blue, 460 fin, 161 humpback, two sei and six sperm) despite the use of

[7]KCCWM, EP, Anders Ellefsen to Atlantic Whaling and Manufacturing Company, 2 May 1905. Ellefsen obtained additional income from the sale of these processed carcasses.

[8]*Evening Herald* (St. John's), 12, 13, 15, 20 and 24 June; 5, 10 and 11 July; and 2 August 1905. Fishermen, for example, reported "that never before were whales seen in such numbers along the south coast, whole schools of them being passed daily." *Humber* "was out from Aquaforte for two days and got back...without a fish. She did not see one on the cruise."

[9]*Ibid.*, 6 and 14 October 1905; and Newfoundland, Department of Fisheries, *Annual Report* (St. John's, 1905), 150. This was confirmed by KCCWM, EP, Letter Book, 18 January 1906. Detailed monthly statistics kept by the company show that ninety-two whales (thirty blue, twenty-eight fin, and thirty-four humpbacks) were killed between April and October 1905.

five new catchers. The Newfoundland Department of Fisheries considered this "a very large deficiency, and demonstrates the great scarcity during the past year."[10]

Whalebone and oil prices also declined, with best quality Grades 0 and 1 oil going from £21 a ton in 1901 to £16 a ton in 1904. Several Newfoundland whaling companies also were forced to sell at £14 10s. in 1905.[11] This price decline had especially important implications for the Aquaforte station, since although Anders Ellefsen was the local operations manager, profit margins and cash flow were controlled from Norway by his father and other major shareholders. This led Anders to plead for easier access to operating capital at the beginning of the 1905 season, including asking Ragnvald Berg to intercede since "it's terrible when your pockets are empty."[12] At the same time, he requested his father to send NOK 10,000 "by June 15th, as there are 3-month bills of acceptance that fall due on that date, so be sure that the money is sent early enough to arrive here on time."[13]

The 1905 *Annual Report* of the Newfoundland Department of Fisheries well described the problems facing the Ellefsens and other whaling investors at the beginning of the 1906 season:

> The price of produce was again low during the year, whale oil only averaged £14 per ton, and finners were lower than even before, so that the few companies which were successful as regards the catch scarcely paid running expenses. The catch of most of the companies was a comparative failure, and the result in several cases has been the winding up of the companies, with the loss of their entire paid up capital...The expenses of

---

[10]Newfoundland, Department of Fisheries, *Annual Report* (1905), 150.

[11]J.N. Tønnessen and A.O. Johnsen, *The History of Modern Whaling* (Berkeley, 1982), 107.

[12]KCCWM, EP, Anders Ellefsen to R. Berg, 21 April 1905. He was forced to buy barrels on "3-month accepts, I have only paid cash for about 300 out of the total of approximately 1500 barrels. About $800.00 in the bank in addition I am expecting roughly $900.00 from Philadelphia and about $300.00 in petty cash. I would have been able to pay cash for more barrels as you can see but was reluctant to make myself short of cash and then have to telegraph too early after money. If only you were able to sell the oil as soon as it arrived so that we could get in some money."

[13]*Ibid.*, Anders Ellefsen to Andreas Ellefsen, 21 April 1905. His position is perhaps best indicated by the fact that in the same letter he explained that he had just purchased a typewriter "from my own money of course, couldn't be that extravagant with the registered company's money."

carrying on this business are very heavy, and like all other fisheries the results are very fluctuating and uncertain.[14]

Table 7
Newfoundland and Labrador Whaling Production, 1906-1907

	Whales		Oil (Gal)		Guano (Tons)		Bone (Tons)	
Factory	**1906**	**1907**	**1906**	**1907**	**1906**	**1907**	**1906**	**1907**
Balaena	24	32	59,392	33,831	212	-	103	4
Snook's Arm	*	75	-	63,024	-	-	-	7
Chaleur Bay	58	37	60,178	36,546	188	-	113	5
Rose-Au-Rue	67	84	44,653	63,720	182	-	106	82.75
Beaverton	24	42	31,006	46,181	95	-	50	7
St. Lawrence	21	30	20,520	26,422	78	93.5	35	49
Cape Broyle	3	2	2,898	2,436	9	-	7	0.25
Cape Charles	25	31	16,640	23,814	-	19	62	3
Dublin Cove	27	27	30,080	25,956	62	70	50	50
St. Mary's	4	*	3,200	-	13	-	6	-
Aquaforte	18+	3	13,824	2,495	-	-	15	0.25
Safe Harbour	21	*	28,300	-	73	-	32	-
Harbour Grace	17	*	22,680	-	65	-	16	-
Hawke's Bay	40	31	83,160	36,000	146	119	69	55
Hawke Harbour	60	63	75,600	73,250	130	125	52	200
Trinity	*	24	-	19,600	-	25	-	40
Totals	**409**	**481**	**492,131**	**453,275**	**1,253**	**451.50**	**718**	**503.25**

Notes: *Station not operating. +Company records indicate 17 whales killed.

Source: Newfoundland, Department of Fisheries, *Annual Reports* (St. John's, 1906 and 1907).

The disastrous 1905 season now led some investors to express the "fear [that] the fishery may soon become extinct." Further, four fewer factories operated in 1906 (see table 7).[15] Despite the reduced catch effort, low catches early in the season indicated that stocks were now at such low levels that even this rate of participation probably could not be sustained.

[14]Newfoundland, Department of Fisheries, *Annual Report* (1905), 150. For details of whaling company closure and amalgamations at the end of the 1905 season see *Evening Herald* (St. John's), 14 August; 3 October; 16 November; and 20 December 1905; and C.W. Sanger and A.B. Dickinson, "Newfoundland and Labrador Shore-Station Whaling: Overexpansion and Decline, 1905-1917," *The Northern Mariner/Le Marin du nord*, X, No. 4 (2000).

[15]*Evening Herald* (St. John's), 14 August 1905. The stations at Snook's Arm, Trinity, Lark Harbour, and L'Anse-au-Loup did not open in 1906.

The dire straits of the Aquaforte station is illustrated in correspondence with the Colonial Cordage Co. in St. John's. When ordering hemp lines for *Humber*, Ellefsen requested additional time to pay for the shipment. The company, recognizing the perilous state of the industry, agreed to provide Ellefsen with a "note at 4 months."[16] Success early in the season, it was hoped, would provide oil and whale bone, and thus ease further access to credit. This did not occur.

Although the first whale taken in the previous two seasons had been caught on 18 April, *Humber* was still in St. John's taking on supplies on that same date in 1906.[17] Not only was the opening of the season delayed but whales were scarce when they finally appeared. *Humber* therefore only killed two by the early part of May and had just "six fish to her credit" by 29 June, usually the most productive months for the station. Most were also taken on longer, more expensive voyages. Earlier in the month, for example, *Humber* had been "in Placentia Bay all the week and got nothing." Other companies with stations on the south coast were doing just as poorly, causing the St. John's *Evening Herald* to report that "the past week [11-17 June] is regarded as the best week of the year for whaling, and it is [sic] utter failure shows better than anything else the collapse of the industry."[18]

Since his workers now were seriously underemployed, Anders Ellefsen attempted to generate additional income to offset losses by having "20 men engaged at the cod fishery there, and they are doing well having 20 qtls. ashore to date." Conditions did not improve in July, and again the whaling crew went cod fishing. *Humber*, for example, had returned to Aquaforte on 24 July, having been "out all yesterday, but did not see a fish. Up to date she has but seven fish, and very poor prospects for the balance of the season." Only four were processed at Aquaforte in August.[19]

---

[16]KCCWM, EP, Colonial Cordage Co. to Anders Ellefsen, 19 March 1906.

[17]*Evening Herald* (St. John's), 17 April 1906. Another indication of the dismal prospects for the 1906 season is provided in a letter received by Anders Ellefsen from a cooper he had previously employed. He ended by expressing the "hope you have some whales by this time [March]. I would have sent before only I heard there was nothing doing." KCCWM, EP, John Carter to Anders Ellefsen, 7 March 1906.

[18]*Evening Herald* (St. John's), 8 May; and 18 and 29 June 1906.

[19]*Ibid.*, 29 June; 14 and 25 July; and 4 September 1906; and KCCWM, EP, John Maclellan to Anders Ellefsen, 6 July 1906. A further indication of the dismal state of the industry, and Ellefsen's operation in particular, was provided in a comment by John Maclellen, Ellefsen's Glasgow oil and bone agent: "I am sorry the fishery has been so poor this season." Further, the *Evening Herald* reported that "the whale fishery is a blank."

The rapidly deteriorating industry seriously affected not only Ellefsen and the other whaling companies but also caused concern in associated industries. When Ellefsen cancelled an order in early August with the Colonial Cordage Co. for whale lines, Managing Director J.H. Munroe replied that "we are extremely sorry to hear that your heavy investment in plant...is not likely to be renumerative this year. We hope that the Fall fishery will make a saving voyage, and that you may have a bumper next year."[20]

Conditions did not improve by the end of September, and the Ellefsens and their Norwegian investors could no longer sustain losses of this magnitude. As early as mid-August, Ragnvald Berg wrote to Anders that:

> Since the whaling is still extremely poor, we must seriously consider withdrawing from Newfoundland...We must now soon have a board meeting...and take these matters under consideration...This is a sorry state of affairs, but there is no other option – we must quit – it is costing money every day and no catch...After the meeting, you will be given clear instructions as to what should be done...I have had enough of this, so I am not able to take part in anything that requires money, I am sorry to say. I think we must just stay calm and save as much as possible of the pieces.

"Clear instructions" included plans to sell or charter *Humber*; ascertaining if the government continued to require an annual licence fee in view of the poor results and imminent bankruptcies; and determining if licences would be required if stocks later revived. Berg also advised that the Board of Directors was seriously considering selling the station: "You had best *prepare* to sell all assets that are saleable – the buildings should probably also be sold. In short, *everything* that is not absolutely necessary for the re-opening of the firm, if so should happen."[21]

Anders and several of the other owners petitioned the Newfoundland government at the end of the 1905 season to end "the $1500 fee for each factory and the coal tax." But the Premier and Executive Council "decided that it could

---

[20]*Evening Herald* (St. John's), 3 September 1906; and KCCWM, EP, Anders Ellefsen to Colonial Cordage Co., 7 August 1906; and J.H. Munroe to Anders Ellefsen, 10 August 1906.

[21]*Evening Herald* (St. John's), 20 September 1906; and KCCWM, EP, R. Berg to Anders Ellefsen, 17 July 1906. The Aquaforte station, in fact, was closed when the 1906 catch produced "a loss so resounding that it wiped out the profits of the four previous years." Tønnessen and Johnsen, *History*, 108. Quite clearly the Ellefsens had learned from their earlier whaling experiences at Finnmark and Iceland.

not be complied with." despite the closing of four stations and the perilous state of the rest. As the Acting Deputy Minister of Marine and Fisheries stated in his response to Ellefsen's request for tax relief, "[y]ou can close your factory if you wish, but while you hold a licence you have to pay."[22]

Selling the Aquaforte premises would also be difficult, since most other owners were in the same predicament. Nevertheless, there were numerous and often ingenious schemes proposed to salvage whatever was possible from the debacle. Job Brothers, for example, enquired if Anders wished to buy its recently constructed "Carcass Plant" at Aquaforte and thus consolidate his operation. He was, of course, unable to "bid for the guano factory at present [as]...this year is not very encouraging." This reply caused Job to enquire if the Aquaforte station would operate in 1907.[23]

The Ellefsens decided that their only option was to stop whaling in Newfoundland. On 2 November 1906 Andreas sent a telegram to Anders informing him that "we must sell factory as quickly as possible."[24] And in early November the following advertisement appeared in local newspapers:

Whaling Property For Sale

> Tenders will be received for the undersigned...for the purchase of the Land, with Whaling Factory, Dwelling House, Bunk House, Boiler House, Coal Shed, Store, Stable, Forge, Slip, Steamer Slip, Machinery and Plant, and Tools and Utensils of the undersigned at Aquaforte. The land contains about 14 acres.
>
> Anders Ellefsen
> Crosbie Hotel
> [St. John's][25]

---

[22]KCCWM, EP, Anders Ellefsen to Minister of Marine and Fisheries, Newfoundland, 25 August 1906; and W.B. Payne to Anders Ellefsen, 11 September 1906; *Evening Herald* (St. John's), 20 and 22 December 1905; and 19 April 1906. With the industry in serious decline at the end of the 1905 season, however, early in 1906 the government reduced the annual whaling licence fee from $1500 to $750.

[23]See, for example, *Evening Herald*, 28 August; 2, 15 and 24 October; and 7, 10, 15 and 17 November 1906; and KCCWM, EP, Job Brothers to Anders Ellefsen, 23 August 1906; Anders Ellefsen to Job Brothers, 25 August 1906; and Job Brothers to Anders Ellefsen, 27 September 1906.

[24]KCCWM, EP, Andreas Ellefsen to Anders Ellefsen, 2 November 1906.

[25]*Evening Herald* (St. John's), 5 November 1906.

There were no acceptable offers, so the Ellefsen family decided to make a bid of their own through their St. John's agent, C.B. Rankin, at a new auction on 5 December. Not wishing to discourage potential buyers in the interim, Anders instructed auctioneer Thomas Clift "not to give any information to newspapers or others concerning same." The factory and plant, 721 oil casks, and fifteen tons of coal cost them $4113.80.[26]

Since the whaling industries of both British Columbia and Japan were now expanding, there was strong competition for the idle Newfoundland catchers.[27] *Humber* was sold to a new company in Japan following a contact from Robert Shankland and Company, London, who had "direct instructions from our special agents in Japan for the purchase of several whalers. We shall be glad to hear whether you would be disposed to sell your *Humber*...This is special and serious business, and we shall be glad if you will furnish us with full particulars promptly." The Norwegian broker Johan Bryde also used Anders Ellefsen as an agent for the purchase of local catchers for the rapidly-expanding Japanese market: "Have brought *Avalon* from Prowse. Kindly help me get crew everything of Harpoon etc. onboard. Engines ready...Wire also if *Mic Mac* or other whaleable [sic] saleable."[28]

Anders continued to buy catchers from those Newfoundland companies likely to cease operations. Henrik and Arne sailed two to Japan, where they were briefly involved in the development phase of this new industry. Their endurance of the hardships and loneliness of distant relocation was typical both of what was expected of those who worked in the whaling industry and of just how desperate the situation had become at Aquaforte. It also characterized the entrepreneurial spirit which initially brought the Ellefsens to Newfoundland. Ragnvald Berg, in a letter exploring the possibility of Henrik taking *Humber* to Japan, for example,

---

[26]*Ibid.*, 21 November 1906; See, also, "Correspondence: C.B. Rankin to and KCCWM, EP, C.B. Rankin to Anders Ellefsen, 5 December 1906; and Anders Ellefsen to T. Clift, 24 November 1906." By Rankin's account the factory and plant cost $3150; 721 casks, $1000; and fifteen tons of coal, $60. Offset against this amount was $12.00 for advertisements in three newspapers and Cliff's two percent commission, $84.20.

[27]See, for example, C.W. Sanger and A.B. Dickinson, "Newfoundland Involvement in 20th Century Shore-Station Whaling in British Columbia," *Newfoundland Studies*, VII, No. 2 (1991), 97-122; and R.L. Webb, *On the Northwest: Commercial Whaling in the Pacific Northwest, 1790-1967* (Vancouver, 1988). Information on the origins of Japanese whaling is provided in Tønnessen and Johnsen, *History*, 128-146.

[28]*Evening Herald*, 27 September and 5 October 1906; and KCCWM, EP, Robertson Shankland and Co. to Anders Ellefsen, 20 December 1906; and Johan Bryde to Anders Ellefsen, 8 November 1906. After continued crew changes and delays, *Humber* finally left Newfoundland on 4 October 1906.

asked Anders to "say hello to Henrik and Hett [his Newfoundland wife] - I am almost afraid of what the latter has to say about Henrik going to Japan."[29]

The Ellefsens, unable to find a buyer for the Aquaforte station, refinanced it and purchased another catcher in Norway for the 1907 Newfoundland season. *Othar* (see figures 50 and 51) had been built at Oslo in 1885 and was thus fourteen-years older than *Cabot* - the very first steam-powered catcher used in Newfoundland and Labrador. Even though the Japanese had considered it unacceptable, the event was hailed in Newfoundland as an act of "faith" by the Ellefsens "in the future of our whaling industry."[30]

Figure 50: *Othar* on Slipway, Aquaforte (1907-1908)

Source: Kmdr. Chr. Christensen Whaling Museum (KCCWM), Ellefsen Papers (EP).

---

[29]KCCWM, EP, Job Brothers to Anders Ellefsen, 3 November 1906; Arne Ellefsen to Anders Ellefsen, 22 March 1907; Andreas Ellefsen to Anders Ellefsen, 2 October 1906; and R. Berg to Anders Ellefsen, 17 July 1906; J.K. Wogn-Henriksen, "Hvalfangst med rot i Stokke II," in *Stokke: Arv og oett* (Stokke, 1983-1984); and Tønnessen and Johnsen, *History*, 108.

[30]*Othar*: Length, seventy-seven feet; breadth, fifteen feet; and depth, eleven feet; sloop-rigged; gross tonnage 73.88. KCCWM, EP, Anders Ellefsen to Harvey and Company, St. John's, 1 August 1908; Andreas Ellefsen to Anders Ellefsen, 12 December 1906; and *Evening Herald* (St. John's), 26 December 1906.

Figure 51: *Othar* Laid-up for Winter, Aquaforte (1907-1908)

Source: See figure 50.

Unfortunately, continued Ellefsen participation became even less likely when *Othar* did not arrive in St. John's from Tønsberg until 17 August, well past the peak catching period. Whales were "very scarce on the Southern Shore," and *Othar* could only take two by the end of the month. On 5 September it also towed a large fin whale into St. John's because "it was too stormy for her to proceed to the station at Aquaforte." The season was not considered "very successful so far," an understatement in that only three whales in total were processed at Aquaforte in 1907. This was obviously an economic disaster for the company, and *Othar* was laid-up in mid-September at Aquaforte for the winter.[31]

[31]*Evening Herald* (St. John's), 19 and 24 August; and 6 September 1907; and KCCWM, EP, "Letterbook," 12 September 1907.

This speculative and desperate attempt to continue whaling had failed, and the Ellefsens faced further debt.[32] The family therefore decided that it could no longer continue whaling in Newfoundland, and Anders again placed advertisements in local newspapers and contacted other Newfoundland and foreign whaling companies. Job Brothers was given first chance to purchase the station because of its "guano factory there [Aquaforte]." Not surprisingly, no offers were received. The closure of stations at Bay Chaleur, Rose-au-Rue, and Beaverton made it even more difficult for *Othar* and the factory and machinery to be sold. Another approach to Job Bros. was unsuccessful, as were letters to prospective buyers in the United States, South Africa, Nova Scotia and other parts of Canada in which Anders indicated that whales locally, in his opinion, "should come back." He also considered that an additional incentive to purchase his station was its "salvage" potential due to the proximity of Aquaforte to the notorious Cape Race, where "when [a] steamer is out whaling, she has [a] good chance of being first at the scene of the wreck." Only one offer was received, from Brister and Son of Halifax for $600.00; it was refused.[33]

Ellefsen's attempts to sell the station through family connections is shown in a letter to F.W. Bissett and Co. in Halifax, NS:

> I am informed through Messrs Salter's Sons, North Sydney, that you have been writing to him enquiring about my Whaling Plant and steamer. S.S. *OTHAR* is...[full details]...in every way well fitted for whaling...[and]...has been hauled up on the shore every winter for 6 or 7 months and well looked after...everything used by me on board the steamer last year will go with her...[full details of "Machinery, Tanks, etc."]...
> I may sell steamer, factory and buildings for $16,000.00...and if you find the price I mention suitable I would like you to let me know as soon as possible, as there are others enquiring about it...[information on Job's guano plant and availability of a Norwegian "foreman"]...To give you an idea of what such a plant...will cost new, I may mention that

---

[32]In an effort to make ends meet, the Ellefsens had extended their "agricultural" operation, attempting to sell eggs to Water Street stores. KCCWM, EP, C.B. Rankin to Anders Ellefsen. 11 December 1907 and 21 September 1908.

[33]*Ibid.*, Anders Ellefsen to Job Brothers, 2 April 1908; Brister and Son, Halifax, to Anders Ellefsen, 25 June 1908; Anders Ellefsen to Joseph Salters, North Sydney, NS, 9 and 11 July 1908; Anders Ellefsen to P.F. Windsorton, Durham, South Africa, 25 July 1908; and "Letterbook," 14 May 1908; and *Evening Herald* (St. John's), 3 July 1908.

> my Plant alone without Steamer cost $25,000, six years ago. Machinery, Buildings and Steamer will not be sold separate.[34]

The itemized inventories of furniture and utensils for the manager's house (appendix V) and for the buildings, property, machinery and supplies (appendix VI), which were circulated to prospective buyers, provide useful insight on the requirements needed for operating an early whaling station in Newfoundland and Labrador.

Increasingly desperate, Anders again decided to re-advertise the vessel and property locally:

> *WHALING PROPERTY, STEAMER AND*
> *OIL CASKS FOR SALE*
>
> *TENDERS* will be received by the undersigned...for the purchase of the Land with Whaling Factory, Dwelling House, Bunk House, Boiler House, Steamer Slip, Machinery and Plant, Tools and Utensils of the undersigned at Aquaforte, the land contains about 13 acres.
> *TENDERS* will also be received for the Whaling Steamer, *Othar*, and 634 new Oil Casks.
> For particulars apply to Mr. C.B. Rankin, Commercial Chambers, St. John's, or the undersigned
>
> Anders Ellefsen,
> Aquaforte[35]

The advertisement was periodically carried by all St. John's newspapers, but did not result in an acceptable offer. As winter approached, and anxious to return home, Anders Ellefsen somewhat despairingly contacted his father in Norway on 10 October 1908: "Cannot obtain an offer. May I leave?" Andreas wired his consent on 18 October, and Anders and Eda, daughter Marie and son Edgar (see figures 52 and 53), the latter born at Aquaforte on 25 February 1906, sailed for Norway from St. John's in SS *Mongolian* on 13 November 1908. Horwood Lumber Co. of St. John's subsequently showed an interest in purchasing *Othar* and the station, perhaps for scrap, but no sale was finalized.

---

[34]KCCWM, EP, Anders Ellefsen to F.W. Bissett and Co., 27 June 1908.

[35]See, for example, *Evening Chronicle* (St. John's), 21, 25 and 28 May 1908.

The Aquaforte property remained under the care of Johan Andreassen, a Norwegian, who was paid $10.75 per month by the Ellefsens.[36]

Figure 52: Anders Ellefsen's Family, Aquaforte (1908). Front: Unidentified friend (left), Edgar (centre), Marie. Rear: Eda (Melvin) and Anders.

Source: See figure 50.

The station was finally sold in 1912 to A/S Kosmos, which was then operating three large combined whaling-sealing vessels in the Northwest Atlantic. Company owner Christian Christensen of Sandefjord, a "leading personality in this significant phase of modern whaling," paid $8000.00 (US) for "The Aquaforte Whaling and Trading Company Ltd.," thus officially severing the connection between the Ellefsens and Newfoundland.[37]

---

[36]*Ibid.*, Anders Ellefsen to Andreas Ellefsen, 10 October 1908; and Anders Ellefsen to Andreas Ellefsen, 20 October 1908; Bruce Melvin, personal communication, 1993; and *Evening Herald* (St. John's) 13 November 1908.

[37]"Christensen, C.L., In Memoriam," *Norsk-Hvalfangste Tidende* (1950), 7-9; Tønnessen and Johnsen, *History*, 96; KCCWM, EP, "Agreement: A/S Kosmos and Anders Ellefsen," 18 August 1911; Newfoundland, Office of the Registrar, Joint Stock

Figure 53: Anders Ellefsen's Newfoundland-born Children, Aquaforte (1908). Unidentified friend (left), Edgar (centre), and Marie.

Source: See figure 50.

A/S Kosmos intended Aquaforte to be the base for *Lloydsen* (Capt. J. Antonisen), which had previously taken part in the 1912 Newfoundland spring seal fishery. While still "at the ice," a crew of Norwegians "began fitting up the premises without delay" in order to prepare *Lloydsen* for whaling off Newfoundland or in Davis Strait:

> The whaler *Lloydsen* sails today to begin the season's work...if the fish are found plentiful near the coast, she will bring her catch to the Aquaforte factory, otherwise she will go north and manufacture the oil on board, being fitted up with tanks and necessary appliances for the purpose. From here [Newfoundland] she will take twelve men...most of whom are familiar

---

Companies, Reg. No. 200; and *Evening Herald* (St. John's), 15 March 1912.

> with the work, having been previously engaged at the different factories.[38]

Lloydsen landed twenty-three whales at Aquaforte before going north to hunt in Greenland waters, "where she took another 15 blue, 5 humpback and 3 fin whales." No further whaling took place from Aquaforte, until rising postwar oil prices caused a sporadic resumption of activities there and at four other locations from about 1923.[39]

---

[38]*Evening Herald*, 15 March and 3 June 1912.

[39]*Ibid.*, 2 September 1912; 4 January 1913; and 13 February 1923; and C.W. Sanger and A.B. Dickinson, "Renewal of Newfoundland and Labrador Shore-Station Whaling, 1918-1936," *International Journal of Maritime History*, VII, No. 1 (1995).

CONCLUDING REMARKS

The Ellefsen family of Stokke, Norway, were compatriots of Svend Foyn, generally regarded as the pre-eminent person in the development of the modern whaling industry. Its members also played an important pioneering role in the evolution of this new phase of whaling, with Andreas and his brother Hans being particularly significant investors in the industry from stations along the Finnmark coast and later in Iceland. Modern commercial whaling expanded globally from its Norwegian origins, as stocks were over-exploited and new ones identified. The Ellefsens continued their involvement in all aspects of this expanding venture as investors, station owners, factory managers, flensing and processing experts, vessel masters, and gunners.

The impact of the Ellefsen family also extended to the developing shore-station industry in Newfoundland and Labrador, where the first factory was established at Snook's Arm in 1898 as a joint venture between local and Norwegian investors. The success of these initial whaling enterprises in making profits by the extraction of oil from whales captured as they migrated along the coast of the island provided the stimulus for the Ellefsens to move into this expanding new trade as their Finnmark and Icelandic operations became less profitable. They constructed a station south of St. John's at Aquaforte, which subsequently became the only totally Norwegian-owned whaling facility on the island. Those family members who came to Newfoundland soon became highly influential in the continued development and operation of the growing local industry, including by serving in important managerial and service positions onshore and at sea for several other companies.

Rapid expansion of the numbers of stations and catchers in the search for quick profits soon led to over-exploitation, which could not be prevented by the ineffective management regulations introduced by the government of Newfoundland and Labrador. Consequently, by 1905 migrating stocks were reduced to the point where most companies and stations were unable to cover operating costs. Catches were much reduced, the industry went into decline, and stations were shut down. Aquaforte was not spared and closed in 1908. The family retained possession of the factory and other property until 1912, when they were sold to another important Norwegian whaling entrepreneur, Christian Christensen of Sandefjord.

The story of the rise and fall of whaling from Aquaforte encapsulates similar changes in fortune that affected all participants in this first cyclical phase of modern shore-station whaling in Newfoundland and Labrador, which came to an end in 1916 with the closure of the station at Hawke Harbour, Labrador. This, however, did not signal the end of whaling in Newfoundland and Labrador, or for the Ellefsen family.

The closing of shore-stations gave whale stocks a temporary respite from hunting, perhaps allowed some slight recovery, and provided the opportunity for periodic renewal of the industry. As a result, new investors entered whaling, particularly as demand and prices for whale oil increased. Consequently, the industry went through four further well-defined cyclical phases until it finally was ended by a government of Canada decree in 1972. Norwegian involvement continued until the end of the industry, as it did elsewhere in global shore-station and pelagic whaling. Without the involvement of Norwegians in the Newfoundland and Labrador industry, it is unlikely that sufficient expertise would have been available to allow this often locally-significant component of the marine resource-dependent economy of Newfoundland and Labrador to continue. This Norwegian presence still occurs, but now in a different area of marine resource utilization, namely the exploration, drilling and production of offshore oil on the Grand Banks off Newfoundland. This region of highly-productive ocean not only supported much of the sporadic local whaling but also the cod fishery, which was the backbone of local economy. Large numbers of whales migrated through these waters in search of food and continue to do so in seemingly larger numbers every year. As a result, another whale-based industry has arisen around Aquaforte: whale-watching and related tourism. The residents of Newfoundland and Labrador thus continue to benefit from their relationship with both Norwegians and whales, as they did when the Ellefsen family and the whales they caught made an important contribution to the economy of rural Newfoundland.

The Ellefsen family continued its involvement in whaling after closing its Newfoundland operation, with Henrik in Japan and South Georgia, Arne in Japan, and Hans and his son Carl at Saldanha Bay, South Africa. Anders returned to Skjærsnes and in 1909 opened Ellefsen's Tenskruefabrikk to make harpoon time-delay fuses following Norwegian patents granted him in 1919 and 1931 (see figure 54). He also farmed and was mayor of Stokke between 1932 and 1941 and again in 1945, when the Nazis replaced him with a local sympathizer. A German plan to manufacture mines at the Ellefsen factory failed when Edgar Ellefsen helped the Norwegian resistance remove a critical piece of machinery. The factory at Skjærsnes is now working again, this time as a machine shop run by a local resident to turn out various small brass fittings.

Anders Ellefsen died on 14 December 1953, followed by his wife Eda (Melvin) on 23 January 1956. She never returned to her native Newfoundland after leaving Aquaforte in 1908. Marie (Kiserud), one of three children of the Anders Ellefsen/Eda Melvin marriage (see figure 55), and the oldest of the two born at Aquaforte, resided in Bygdøy, Oslo, until her death in 1995, a fine result of early harmonious Newfoundland-Norwegian relationships (see figure 56). Her two sons, Børre and Egil Kiserud, together with Egil's son Anders, re-established contact with their grandmother's roots when they visited Newfoundland and their Melvin relatives in 1990.

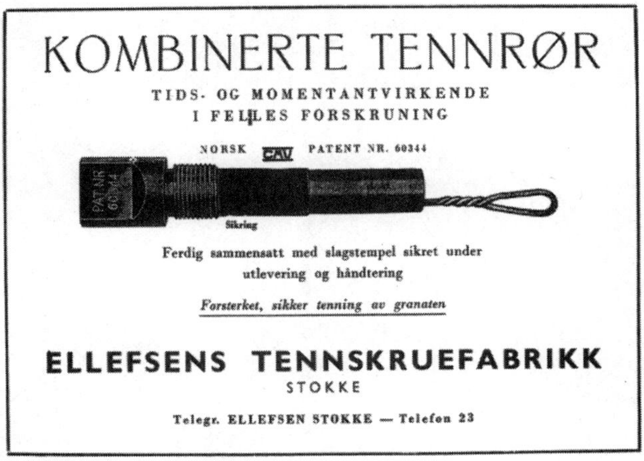

Figure 54: Advertisement for Ellefsen's Time-Delay Fuses

Source: Kmdr. Chr. Christensen Whaling Museum (KCCWM), Ellefsen Papers (EP).

Figure 55: Ellefsen family in Norway, 1931. (L-R): Eda, Marie, Edgar, Evelyn (born in Stokke) and Anders.

Source: KCCWM, EP, courtesy of A. Elton-Morris, St. John's.

Figure 56: Twenty-four Friends and Family Celebrate Ninetieth Birthday of Marie (centre), Oslo, 11 January 1994

Source: Courtesy of Børre and Egil Kiserud.

Newfoundland and Labrador Whaling, 1908-1972

This decline of the first phase of modern shore-station whaling in Newfoundland and Labrador had a significant impact on the development of operations elsewhere in British Columbia. Dr. Ludwig Rissmuler in particular foresaw the new opportunities and relocated to Victoria in 1905 to become an active partner in the newly- established Pacific Whaling Co. Ltd. Connections established in Newfoundland during ownership of the St. Lawrence Whaling Co. Ltd. and the siting of guano factories adjacent to other whaling stations provided him with access to a supply of managers and skilled workers for the new industry. These men and their families moved to British Columbia and had a lasting and important impact on the development and continuation of the Canadian west coast whaling industry.[1]

[1]For greater details on the cyclical nature of Newfoundland and Labrador whaling after the close of the Aquaforte factory, see A.B. Dickinson and C.W. Sanger, "Newfoundland and Labrador Shore-Station Whaling: The Third Major Phase, 1937-1951," *International Journal of Maritime History*, XI, No. 1 (June 1999); Dickinson and Sanger, "Newfoundland and Labrador Shore-Station Whaling: The Final Demise, 1951-

Despite two unsuccessful attempts at revitalizing the local industry in 1918 and 1919, Newfoundland and Labrador shore-station whaling languished until 1923 when a Norwegian whaling master, Captain Amund Anonsen, formed the Newfoundland Whaling Co. Ltd. with Newfoundland and Norwegian capital. By 1925, Anonsen was harvesting whales off the north and south coasts from two of the older stations, Hawke Harbour on the southern coast of Labrador and Rose-au-Rue in Placentia Bay. Catches again declined after 1928, and profits were also significantly affected by lower whale oil prices and reduced demand due to global oversupply and the increased availability of unsaturated vegetable oils.

Following the lead of the Newfoundland Whaling Co. Ltd., Svend Foynd Bruun and Anton von der Lippe of Tønsberg, Norway, also went into partnership with Job Brothers of St. John's in late 1926 to incorporate the British Norwegian Whaling Co. Ltd. Instead of reopening one of the older factories, a new plant was built at Watering Cove, Big Grady Island, Labrador, to take advantage of the more successful northern whaling grounds. But the Grady factory was located in a "poor harbour, the wharf having to be constructed distant from the plant and connected by a railway" and thus had only limited success. Both companies ceased operations during 1931-1932.[2]

The Newfoundland Whaling Co. Ltd. was restructured in 1933 and resumed hunting the following year from Hawke Harbour and Rose-au-Rue. Also beset by limited resources and marginal success, the company formed an alliance with the major global whaling enterprise, Christian Salvesen and Co. Ltd. of Leith, Scotland. Its subsidiary, the Polar Whaling Co. Ltd., was set up to carry out Newfoundland operations and began whaling in 1937. Before the season came to a close, 483 whales were taken, "one of the best [records] in the history of

---

1972," *The Northern Mariner/Le Marin du Nord*, IX, 3 (1999); Sanger and Dickinson, "Newfoundland and Labrador Shore-Station Whaling: Overexpansion and Decline, 1905-1917." *The Northern Mariner/Le Marin du nord*, X, No. 4 (October 2000), 13-38;" Sanger and Dickinson, "Whaling," in C.F. Poole (ed.). *Encyclopedia of Newfoundland and Labrador* (5 vols., St. John's, 1981-1994), V, 546-551; and Sanger and Dickinson, "Renewal of Newfoundland and Labrador Shore-Station Whaling, 1918-1936," *International Journal of Maritime History*, VII, No. 1 (June 1995). Fuller accounts of Newfoundland involvement in the establishment of British Columbia commercial whaling are provided by Sanger and Dickinson, "Newfoundland Involvement in 20th Century Shore-Station Whaling in British Columbia," *Newfoundland Studies*, VII, No. 2 (1991), 97-122; and Sanger and Dickinson, "*They Were Clannish as Hell": Origins of Shore-Station Whaling in British Columbia - The Newfoundland Factor* (Halifax, 1991).

[2]D.E. Sergeant, "Whaling in Newfoundland and Labrador Waters," *Norsk Hvalfangst-Tidende*, XLII, No. 12 (1953).

whaling in this country."[3] Another oil market collapse, and internal disagreements between Salvesen and Anonsen as to the management of the Newfoundland and Labrador operation, brought the industry to another halt in 1938.

Hunting resumed on a reduced scale in 1939, Salvesen activities in Newfoundland and Labrador were further diminished in 1940 when the Ministry of Shipping in London requisitioned for war duty those Polar Whaling Co. Ltd. catchers considered suitable for coastal defence of Britain. Despite the loss of its most efficient vessels, the company conducted limited whaling from Hawke Harbour in 1940, but then ceased operations until 1943. The Rose-au-Rue station, meanwhile, was purchased by a new company, Marine Oils Ltd., and processed a small number of whales captured in Placentia Bay during 1941-1942.[4]

In the latter half of the 1940s, Newfoundland and Labrador shore-station whaling enjoyed a marked revival. Increased postwar demand for whale products, a series of poor oil seed harvests, and a consequent rise in oil prices, attracted further attention from local investors. New stations were therefore opened at Williamsport in 1945 and Dildo two years later. Local whale stocks continued to be unable to sustain this level of renewal, and by the early 1950s the industry was once again in decline.[5] Salvesen severed all ties with Newfoundland and Labrador shore-station whaling in 1956, and attempts by others to continue the trade were further discouraged by significantly reduced global demand for whale oil. Limited attempts to resume large-scale whaling during the 1950s and early 1960s were linked closely with fur farming and the catching of small whale species. The Arctic Fishery Products Co. Ltd. at Dildo was built primarily to process pothead and minke whale meat and oil, and to help ease meat shortages on local mink and fox farms.[6]

The final resurgence of Newfoundland and Labrador commercial whaling began in 1966, when Arctic Fishery Products Co. Ltd. began using the Dildo factory to process meat for human consumption in Japan. The Japanese became directly involved in local whaling when Taiyo Gyogyo K.K. entered into a partnership with Arctic Fishery Products to form the Atlantic Whaling Co. Ltd. Operating from the refurbished Williamsport station, the new company processed 1248 fin, humpback and sei whales between 1967 and 1972 for 33,063 barrels of oil.

---

[3]*Evening Telegram*, 26 October 1937. For additional information on Salvesen involvement in local whaling see, W. Vamplew, *Salvesen of Leith* (Edinburgh, 1975).

[4]*Evening Telegram*, 12 October 1942.

[5]*Ibid.*, 25 May 1948.

[6]See, for example, *ibid.*, 22 January 1949; 20 April 1955; and 24 July 1964.

Worldwide concern about the status of whale stocks led Canada to place a moratorium on commercial whaling on 21 December 1972, "not merely on economic grounds but primarily in order to make its contribution to saving world whale stocks."[7] The Dildo and Williamsport stations therefore closed, signalling the end of seventy-five years of modern commercial whaling in Newfoundland and Labrador coastal waters. Altogether, twenty-one factories, strategically located close to stocks migrating around the island and southern Labrador, had processed more than 19,000 whales. The activities described for the Ellefsen operations at Aquaforte represent the only reasonably complete example of how such whaling was implemented.

[7]J.N. Tønnessen and A.O. Johnsen, *The History of Modern Whaling* (Berkeley, 1982).

BIBLIOGRAPHY

Allen, K.R. "Catch Per Unit Effort of Northwest Atlantic Fin Whale Stocks." International Whaling Commission. *Twenty-third Report*. Cambridge, 1973, annex H.

Allen, K.R. *Conservation and Management of Whales*. London, 1980.

"An Act to Regulate the Whaling Industry." *Acts of the General Assembly of Newfoundland, Passed in the 2nd Year of the Reign of His Majesty King Edward V*. St. John's, 1902.

Ashley, C.W., *The Yankee Whaler*. London, 1938.

Australia. *Whales and Whaling; Independent Inquiry Report*. 2 vols. Canberra, 1979.

Baker, M.; Dickinson, A.B.; and Sanger, C.W. "Adolph Nielsen: Norwegian Influence on Newfoundland Fisheries in the Late 19th-Early 20th Century." *Newfoundland Quarterly*, LXXXVII, No. 2 (1992), pp. 25-35.

Barkham, S. "The Basques: Filling a Gap in Our History Between Jacques Cartier and Champlain." *Canadian Geographical Journal*, XCVI, No. 1 (1978), pp. 8-19.

Barkham, S. "The Identification of Labrador Ports in Spanish 16th Century Documents." *Canadian Cartographer*, XIV, No. 1 (1977), pp. 1-9.

Basberg, B.L. "Survival Against all Odds? Shore Station Whaling at South Georgia in the Pelagic Era, 1925-1960." In Basberg, B.L.; Ringstad, J.E.; and Wexelsen, E. (eds.). *Whaling and History: Perspectives on the Evolution of the Industry*. Sandefjord, 1993.

Bockstoce, J. "History of Commercial Whaling in Arctic Alaska." *Alaska Geographic*, V, No. 4 (1978).

Bockstoce, J.R. *Steam Whaling in the Western Arctic*. New Bedford, 1977.

Budker, P. *Whales and Whaling*. London, 1958.

Burton, R. *The Life and Death of Whales*. London, 1973.

"Christensen, C.L., In Memoriam." *Norsk Hvalfangst Tidende* (1950), pp. 7-9.

Christensen, T. "Preliminary Report on the Norwegian Fishery for Small Whales: Expansion of Norwegian Whaling to Arctic and Northwest Atlantic Waters and Norwegian Investigation of the Biology of Small Whales." *Journal of the Fisheries Research Board of Canada*, XXXII, No. 7 (1975), pp. 1083-1094.

Clarke, A.H. "History and Present Condition of the Fishery." In Goode, G.B. (ed.). *The Fisheries and Fishery Industries of the United States*. 4 vols. Washington, DC, 1887.

Conway, M. *No Man's Land: A History of Spitsbergen from its Discovery in 1596 to the Beginning of Scientific Exploration of the Country*. Cambridge, 1906.

Conway, M. (ed.). *Early Dutch and English Voyages to Spitsbergen in the Seventeenth Century*. London, 1904.

De Jong, C. *A Short History of Old Dutch Whaling*. Pretoria, 1978.

Dickinson, A.B. and Sanger, C.W. "Modern Shore-based Whaling in Newfoundland and Labrador: Expansion and Consolidation, 1898-1902." *International Journal of Maritime History*, II, No. 1 (1990), pp. 83-116.

Dickinson, A.B. and Sanger, C.W. "Modern Shore-Station Whaling in Newfoundland and Labrador: The Peak Season, 1904." *International Journal of Maritime History*, V, No. 1 (1993), pp. 127-154.

Dickinson, A.B. and Sanger, C.W. "Newfoundland and Labrador Shore-Station Whaling: The Third Major Phase, 1937-1951." *International Journal of Maritime History*, XI, No. 1 (1999), pp. 101-116.

Dickinson, A.B. and Sanger, C.W. "Newfoundland and Labrador Shore-Station Whaling: The Final Demise, 1951-1972." *The Northern Mariner/Le Marin du Nord*, IX, No. 3 (July 1999), pp. 39-52.

Douglas, W.A. and Bilbas, J. *Amerikanuak: Basques in the New World*. Reno, NV, 1975.

Dundee Advertiser, various years.

Edmundson, G. *Anglo-Dutch Rivalry During the First Half of the Seventeenth Century*. Oxford, 1911.

Ellefsen Papers. Kmdr. Chr. Christensen Whaling Museum, Sandefjord, Norway.

Ellis, R.E. *Men and Whales*. New York, 1991.

Evening Chronicle (St. John's), various years.

Evening Herald (St. John's), various years.

Evening Telegram (St. John's), various years.

Falnes, J. *Løst Krutt og Knallpester: Hvalfangstbestyren som ble fabrikkeier i Stokke*. Sandefjord, 1999.

Gaskin, D.E., *Whales, Dolphins and Seals, with Special Reference to the New Zealand Region*. London, 1972.

Goddard, J.M. "The Rismuller Factor in North American Shore Whaling, 1900-1912: The Atlantic Years." *International Journal of Maritime History*, V, No. 2 (1993), pp. 131-155.

Gulland, J.A. and Boerema, L.K. "Scientific Advice on Catch Levels." International Whaling Commission. *Twenty-third Report*. Cambridge, 1973, annex S.

Harmer, S.F. "History of Whaling." *Proceedings of the Linnean Society of London*, CXL (November 1927-May 1928), pp. 51-95.

Hay, K. *Guide to Identification of Whales in Newfoundland Waters*. St. John's, 1979.

Hay, K.A. "Status of the Humpback Whale, *Megaptera novaeangliae*, in Canada." *Canadian Field-Naturalist*, XCIX, No. 3 (1985), pp. 425-432.

Headland, R.K. *The Island of South Georgia*. Cambridge, 1984.

Hohman, E.P. *The American Whaleman: A Study of Life and Labor in the Whaling Industry*. New York, 1928.

International Whaling Statistics. Oslo, 1930-1980.

Job, W.C., "Whaling Notes Compiled by William C. Job on a Visit to Norway, June 1903." Typescript. Public Archives of Newfoundland and Labrador, Job Papers.

Job Papers. Memorial University of Newfoundland.

Lein, J.; Fawcett, L.; and Staniforth, S. *Wet and Fat: Whales and Seals of Newfoundland and Labrador*. St. John's, 1985.

Macy, Obed. *The History of Nantucket: Being a Compendious Account of the First Settlement of the Island by the English, Together with the Rise and Progress of the Whale Fishery; and Other Historical Facts Relative to Said Island and Its Inhabitants*. Boston, 1835.

Mackintosh, N.A. *The Stocks of Whales*. London, 1965.

Markham, C., "On the Whale-Fishery of the Basque Provinces of Spain." *Proceedings of the Zoological Society of London* (13 December 1881), pp. 969-976.

Meredith, G.R. and Campbell, R.R. "Status of the Fin Whale, *Balaenoptera physalus*, in Canada." *Canadian Field-Naturalist*, CII, No. 2 (1988), pp. 351-368.

Milinkovitch, M.C., *et al.* "Revised Phylogeny of Whales Suggested by Mitochondrial Ribosomal DNA Sequences." *Nature*, CCCLXI (1993), pp. 346-348.

Mitchell, E.D. "Assessments of Northwest Atlantic Fin Whale Stocks." International Whaling Commission. *Twenty-second Report*. Cambridge, 1972, annex L.

Mitchell, E.D. "Draft Report on Humpback Whales Taken under Special Scientific Permit by Eastern Canadian Land Stations, 1969-71." International Whaling Commission. *Twenty-third Report*. Cambridge, 1973, annex M.

Mitchell, E.D. "Present Status of Northwest Atlantic Fin and Other Whale Stocks." In W.E. Schevill (ed.). *The Whale Problem: A Status Report*. Cambridge, MA, 1974.

Mitchell, E.D. and Reeves, R.R. "Catch History, Abundance, and Present Status of Northwest Atlantic Humpback Whales." In Tillman, M.F. and Donovan, G.P. *Historical Whaling Records*. Cambridge, 1983.

Newfoundland. Department of Fisheries. *Annual Report*. St. John's, 1903.

Newfoundland. Department of Fisheries. *Annual Report*. St. John's, 1904.

Newfoundland. Department of Fisheries. *Annual Report*. St. John's, 1905.

Newfoundland, Department of Fisheries. *Annual Report*. St. John's, 1906.

Newfoundland. Department of Fisheries. *Annual Report*. St. John's, 1907.

Newfoundland. Registry of Companies.

Newfoundland. Office of the Registrar, Joint Stock Companies, Reg. No. 200.

Niemi, E. "Modern Whaling on the Norwegian Arctic Coast - Origin, Development and the Local Society." In Basberg, B.L.; Ringstad, J.E.; and Wexelsen, E. (Eds.), *Whaling and History. Perspectives on the Evolution of the Industry*. Sandefjord, 1993.

Norsk Hvalfangst Tidende. various issues.

"A Note on Baleen Whale Stocks of the Northwest Atlantic." International Whaling Commission. *Twentieth Report*. Cambridge, 1970, annex I.

Novacek, M. "Genes Tell a New Whale Tale." *Nature*, CCCLXI (1993), pp. 298-299.

Purchas, S. *Hakluytus Posthumus, or Purchas, His Pilgrimes, Contayning a History of the World in Sea Voyages and Lande Travells by Englishmen and Others*. London, 1625; reprint, 20 vols., Glasgow, 1905-1907.

Sanger, C.W. "The 19th Century Newfoundland Seal Fishery and The Influence of Scottish Whalemen." *Polar Record*, XX (1980), pp. 231-352.

Sanger, C.W. "The Origins of the Scottish Northern Whale Fishery." Unpublished PhD thesis, University of Dundee, 1985.

Sanger, C.W. and Dickinson, A.B. "The Origins of Modern Shore-Based Whaling in Newfoundland and Labrador: The Cabot Steam Whaling Co. Ltd., 1896-98." *International Journal of Maritime History*, I, No. 1 (June 1989), pp. 129-157.

Sanger, C.W. and Dickinson, A.B. *"They Were Clannish as Hell": Origins of Shore-Station Whaling in British Columbia - The Newfoundland Factor*. Halifax, 1991.

Sanger, C.W. and Dickinson, A.B. "Newfoundland Involvement in 20th Century Shore-Station Whaling in British Columbia." *Newfoundland Studies*, VII, No. 2 (1991), pp. 97-122.

Sanger, C.W. and Dickinson, A.B. "Expansion of Regulated Modern Shore-Station Whaling in Newfoundland and Labrador, 1902-03." *The Northern Mariner/Le Marin du Nord*, I, No. 2 (April 1991), pp. 1-22.

Sanger, C.W. and Dickinson, A.B. "Modern Shore Station Whaling in Newfoundland and Labrador: The Peak Season, 1904." *International Journal of Maritime History*, V, No. 1 (June 1993), pp. 127-154.

Sanger, C.W. and Dickinson, A.B. "Whaling." In C.F. Poole (ed.). *Encyclopedia of Newfoundland and Labrador*. 5 vols. St. John's, 1981-1994. V, pp. 546-551.

Sanger, C.W. and Dickinson, A.B. "Renewal of Newfoundland and Labrador Shore-Station Whaling, 1918-1936." *International Journal of Maritime History*, VII, No. 1 (June 1995), pp. 83-103.

Sanger, C.W. and Dickinson, A.B. "The Construction and Display of the First Full-Scale Model of a Blue Whale: The Newfoundland Connection." *Acadiensis*, XXVII, No. 1 (1997), pp. 67-84.

Sanger, C.W. and Dickinson, A.B. "Newfoundland and Labrador Shore-Station Whaling: Overexpansion and Decline, 1905-1917." *The Northern Mariner/Le Marin du nord*, X, No. 4 (October 2000), pp. 13-38.

Scammon, C.M. *The Marine Mammals of the North-Western Coast of North America Described and Illustrated, Together with an Account of the American Whale-fishery*. San Francisco, 1874.

Schmitt, F.P.; De Jong, C.; and Winter, F.H. *Thomas Welcome Roys: America's Pioneer of Modern Whaling*, Charlottesville, 1980.

Scoresby, W., Jr. *An Account of the Arctic Regions, with a History and Description of the Northern Whale-Fishery*. 2 vols. Edinburgh, 1820; reprint, Newton Abbot, 1969.

Sergeant, D.E. "Whaling in Newfoundland and Labrador Waters." *Norsk Hvalfangst Tidende*, LXII, No. 12 (1953).

Starbuck, A. *History of the American Whale Fishery from its Earliest Inception to the Year 1876*. 2 vols. Washington, DC, 1877.

Tocque, Philip. *Kaleidoscope Echos: Being Historical, Philosophical, and Theological Sketches, from the Miscellaneous Writings of Philip Tocque*. Toronto, 1895.

Tønnessen, J.N. and Johnsen, A.O. *Den Moderne Hvalfangste Historie, Opprinelse Og Utviking*. 4 vols., Sandefjord, 1959-1970.

Tønnessen, J.N. and Johnsen, A.O. *The History of Modern Whaling*. Berkeley, 1982.

True, F.W. *The Whalebone Whales of the Western North Atlantic Compared with Those Occurring in European Waters, with Some Observations on the Species of the North Pacific*. Washington, DC, 1904.

Tuck, J.A. "Unearthing Red Bay's Whaling History." *National Geographic*, CLXVIII, No. 1 (1985), pp. 50-57.

Tuck, J.A. and Grenier, R. "A 16th-Century Basque Whaling Station in Labrador." *Scientific American*, CCXLV, No. 5 (1981), pp. 180-188.

Vamplew, W. *Salvesen of Leith*. Edinburgh, 1975.

Watson, L. *Sea Guide to Whales of the World*. London, 1981.

Webb, R.L. "Invented Too Late: The Introduction of Steam to the Arctic Whaling Fleet." *American Neptune*, XLIV (1984).

Webb, R.L. *On the Northwest: Commercial Whaling in the Pacific Northwest, 1790-1967*. Vancouver, 1988.

Whitehead, H. "Updated Status of the Humpback Whale, *Megaptera novaeangliae*, in Canada." *Canadian Field-Naturalist*, CI, No. 2 (1987), pp. 284-294.

Wogn-Henriksen, J.K. "Hvalfangst med rot i Stokke II." In *Stokke: Arv og oett*. Stokke, 1983-1984, pp. 19-29.

APPENDICES

Appendix I
Norwegian Whaling Companies With Origins in Stokke

1.	Year	1881 to 1882
	Company	Stokke Interessentskap for Hvalfangst
	Station	Stegelnes, in Bussesund near Vardø
	Catchers	1. *Pasvig*
		2. *Varanger*
	Owner/Manager	Andreas Ellefsen, Stokke
2.	Year	1883 to 1888
	Company	The Skjærsnes Company, Stokke/Kristiania
	Station	Stegelines, in Bussesund near Vardø
	Catchers	1. *Nora*
		2. *Ottar*
	Owner/Manager	Hans Edvard Ellefsen, Stokke and Kristiania
3.	Year	1883 to 1891
	Company	Stokke Partnership for Whaling
	Station	Småstraumen, in Ytre Jarfjord, Pasvik
	Catchers	1. *Pasvig*
		2. *Varanger*
	Owner/Manager	Andreas Ellefsen
4.	Year	1892 to 1901
	Company	Stokke Partnership for Whaling
	Station	Risfjord, by Gamvik
	Catchers	1. *Pasvig*
		2. *Varanger*
	Owner/Manager	Andreas Ellefsen

Source: Kmdr. Chr. Christensens Whaling Museum (KCCWM), Ellefsen Papers (EP).

Appendix II
Icelandic Whaling Companies With Origins in Stokke

1.	Year	1889 to 1901
	Company	Hans E. Ellefsen, Skjærsnes/Kristiania
	Station	Solbakki, at Flateyri in Ørnundarfjødur, Iceland
	Catchers	1. *Nora* 2. *Ottar* 3. *Snorre Sturlasson* 4. *Ørnundarfjødur* 5. *Njåll* 6. *Mosvalla* 7. *Ingolfr* 8. *Pasvig* 9. *Varanger* 10. *Hjorleifr* 11. *Jon Sigurdsson* 12. *Naddadur* 13. *Skarphedin* 14. *Høgni*
	Owner/Manager	Hans Ellefsen. Managers included Jørgen Øre, Teie and Fritjof Ellefsen
2.	Year	1901 to 1911
	Company	Hans E. Ellefsen
	Station	Askneset, in Mjoafjørdur, Iceland
	Catchers	1. *Nora* 2. *Ottar* 3. *Snorre Sturlasson* 4. *Ørnundarfjødur* 5. *Njåll* 6. *Mosvalla* 7. *Ingolfr* 8. *Pasvig* 9. *Varanger* 10. *Hjorleifr* 11. *Jon Sigurdsson* 12. *Naddadur* 13. *Skarphedin* 14. *Høgni*
	Owner/Manager	Hans Ellefsen. Managers included Jørgen Øre, Teie and Fritjof Ellefsen

Source: See appendix I.

Appendix III
Wages and Occupations of Newfoundlanders Employed at Aquaforte Whaling Station, 1902-1903

Name	Occupation	Year	Months	Total wage ($) (including overtime).	Basic
Henry Ryan	Flenser	1902	5	165.75	100 20/month
		1903	5	173.73	105 21/month
Patrick Byrne	Flenser	1902	5	141.18	80 16/month
		1903	5	148.85	80
Peter Croft	Flenser	1902	5	164.77	100 20/month
		1903	5	148.38	80 16/month
Bert Croft	Flenser	1902	5	166.37	100 20/month
		1903		161.68	92.50 18.50/month
Michael Meagher	Flenser	1902	6	141.29	9680
		1903	5	147.44	
James Meagher	Flenser	1902	6	123.93	120
Richard Meagher	Flenser	1902	6	139.46	1201
		1903	5	179.86	20
William Meagher	Flenser	1902	6	135.43	1201
		1903	5	100.55	00
James Meagher	Flenser	1902	6	142.99	120
		1903	5	169.66	100
William Croft	Flenser	1902	6	123.88	120
Patrick Saunders	Flenser	1902	6	124.03	120
		1903	5	100.20	100
Peter Case	Flenser	1902	6	137.72	120
		1903	5	154.00	100
Henry Winsor	Flenser	1902	6	155.28	124 20.67/month
		1903	5	100.40	100
William Croft	Flenser	1902	6	123.44	120
		1903	5	101.80	100
Frederick Ryan	Flenser	1902	6	123.44	120
		1903	5	101.70	100
John Case	Flenser	1902	6	121.89	120
		1903	5	100.00	100
William Keough	Flenser	1902	6	127.84	120
John Healy	Flenser	1902	6	144.23	120
		1903	5	134.98	100
George Johns	Flenser	1902	6	170.16	144 24/month
William Canning	Flenser	1902	6	121.52	120
George Croft	Flenser	1902	6	147.11	144
John Ryan	Flenser	1902	6	161.05	156 26/month
		1903	5	131.10	130 21.67/month
James Winsor	Flenser	1902	6	151.27	144
Alan Graham	Engineer	1902	6	139.41	132 22/month
		1903	5	122.76	120
James Hiscock	Flenser	1902	6	124.18	120
Patrick Welsh	Flenser	1902	6	125.19	120

Name	Occupation	Year	Months	Total wage ($) (including overtime).	Basic
Bernard Dale	Flenser	1902	6	250.15	180 30/month
		1903	5	251.82	150
Frederick Dale	Flenser	1902	6	128.80	120100
		1903	5	100.35	
John Carter	Cooper	1902	6	184.25	180
		1903	-	188.00	-
William Grimes	Blacksmith	1902	4	82.93	73 18.25/month
Bessie Meagher	Housekeeper (?)	1902	-	51.35	- 7/month
		1903	-	4.65	- 7/month
Annie Croft	Cook	1902	-	58.67	- 8/month
		1903	-	5.33	- 8/month
William Croft, Jr.	Flenser	1902	5	130.60	130
John Ryan	Flenser	1902	6	147.45	-
		1903	4	100.00	-

Total Wages: **1902 = $4,677.01**
1903 = $2,927.24

Source: KCCWM, EP, "Employment and Wage Ledger, 1902-1903."

Appendix IV
Wages and Occupations of Norwegians Employed at Aquaforte Whaling Station, 1902

Name	Occupation	Total Wages (Kr.) (Incl. Over time)	Approximate $ (@ 3.70 Kr)
Ole Kristiansen	1st Mate	1331.00	359.73
Oscar Hansen	2nd Mate	258.66	69.91
Anders Simmerlund	--	520.00	140.54
N. Nilsen	Seaman	666.00	180.00
S. Nilsen	Seaman	676.50	182.84
H. Gulliksen	Seaman	704.00	190.27
Anders Andreassen	Skipper	666.00	180.00
Richard Lorentsen	Skipper	704.00	190.27
Henrik Ellefsen	Skipper	2444.27	660.61
Johann Andreassen	Skipper	1270.46	343.37
M. Millelsen	Skipper	2207.07	596.50
Kristian Hansen	Skipper	6017.00	1626.22
Carl Larsen	Blacksmith	385.00	104.05
Anton Sørensen	1st Mate	395.38	106.86
	Totals =	**18,245.34 Kr**	**$4,931.17**

Source: See appendix III.

Appendix V
Manager's House (Furniture and Utensils), Aquaforte Whaling Station, 1908

Items	When bought	At present
1 bureau and stand	$22.35	$11.00
2 bedroom chairs	2	1
1 mattress	6	3
1 spring	5	2.5
1 fancy toilet set (6 pieces)	3.5	2
1 white toilet set	1.65	1
6 chairs	9.6	5
1 arm chair	2.9	1.5
1 extention table	14	8
1 lounge	13	5
1 office chair	8	5
1 Eng. harp lamp	4.6	2
1 office lamp	1.9	1
1 hall lamp	2.95	1.5
1 library lamp	6.75	4
3 kitchen chairs	1.5	0.3
5 cherry curtain poles	1.7	1
6 window poles	1.98	1.2
1 set fire brasses	5	1
1 table bell	0.85	0.4
1 caffee pot	1.4	0.9
1 set carvers	5.5	2.5
1 cruet stand	3.45	0.75
1 salt shake	0.15	0.1
1 doz. table spoons	5.2	3
1 doz. dessert spoons	3.5	2
1 doz. tea spoons	2	1

Items	**When bought**	**At present**
1 doz. table forks	5.2	3
1 doz dessert forks	3.8	2
1 coal vase	2	0.5
1 dinner set (68 pieces)	18	8
1 doz. egg cups	0.4	0.3
1 fancy china milk jug	0.55	0.3
1 fancy china suker basin	0.4	0.25
1 fancy china butter cooler	1	0.5
1 glass dish	0.15	0.1
1 glass dish	0.35	0.25
2 fan. candle sticks	0.36	0.1
2 cut glass decanters	2.8	1.5
1 doz. dessert knives	4.25	2.25
1 doz. table knives	6	3.5
1 mirror	0.3	0.1
1 mirror	0.6	0.4
1 family scale	1.5	1
1 doz. etched wine glasses	1.5	0.5
1 doz. tumblers (etched)	0.95	0.3
2 spring blinds	0.76	0.2
1 doz. spring blinds	6.5	2
1 feather pillow	1.25	0.65
2 feather pillows	3	1.5
1 down quilt	7.5	2
1 pr. blankets	7.75	3
1 pr. blankets	4.8	2
1 quilt	1.55	0.2
50 yds. sheeting	5.25	2.5

Items	**When bought**	**At present**
20 yds. sheeting	9.6	5
6 towels	2.88	1.2
6 pillow slips	1.98	1
1 doz. pillow slips	1.1	0.5
1 £ napkins	2	1
2 table cloths	8.2	4.2
2 table cloths	7	3
1 wadded quilt	3.5	1.5
1 frying pan	0.42	0.2
2 baking pots	0.6	0.4
1 oval boiler	2.3	1
2 jelly shapes	0.6	0.2
1 gall. oil can	0.95	0.45
1 spy glass	20	8
1 barometer	7	3.5
1 gauging rod	1.5	0.5
10 charts	7.5	3
1 box mathematical instruments	1.2	0.5
1 set irons	1.25	0.7
Total	$304.56	$142.40

Note: Estimated value if sold by auction or private sale.

Source: KCCWM, EP, Anders Ellefsen to Messrs F.W. Bissett and Co., Halifax, NS, 27 June 1908.

Appendix VI
Buildings, Machinery, and Equipment, Aquaforte Whaling Station, 1908

Dwelling House - 2 story wooden building, 38 x 28 ft.

Effects

1 no. 2 Champion stove
2 no. 1 Champion stoves
2 no. 13 Peri stoves
1 no. 7 Ideal stove
1 Iron Safe
1 Office Table

Bunk House, 1 story wooden building, 38 x 27 ft.

Effects

1 Kettle
2 Kitchen chairs
2 Boilers
1 Coffee kettle
2 Large frying pans
1 Set tea scales
2 no. 2 Tortoise stoves
1 no. 6 Ideal stove
1 Stove
1 Large caboose
1 Mincing machine

Factory, 2 story wooden building, 54 x 50 ft.

Machinery, tanks, etc. belonging to same
4 Boiling tanks (about) 8 x 16 ft.
4 Cooling tanks (about) 8 x 8 x 10 ft.
4 Blubber tanks (about) 9 x 7 x 7 ft.
1 Cooling tank (about) 16 x 8 x 4 ft.
1 Wooden cooling tank 12 x 6 x 5 ft.
1 Elevator, about 90 ft. long
Shafting and Pulleys
2 Grindstones
1 Set Pulley block
1 Trawly screw jack
1 Double winch
1 Portable Forge
2 Pipe wrenches
2 Pipe cutters
1 no. 3 1/2 Jerrick dies
1 Bell
2 Large Galvanized blocks
1 Anchor, wooden stock
1 Hand Saw
1 Ice Saw
1 Cross cut Saw

2 Tackles
1 White painted boat
1 Boat, flat bottom
5 pr. Oars
1 Steam Engine
1 Slicing Machine
1 Lathe
1 Sawing Machine
10 Wooden buckets
2 Zinc buckets (old)
3 Funnels
2 Oil cans
4 Flencing Knifes

Forge, wooden building

Effects

1 Bellows
1 Anvil
1 Sledge
1 Hammer
1 Ratchet drill
1 Brace
1 Box Stocks & Dies
1 Vise
1 Hand Vise
Drills, chisels and other tools
used for straightening harpoons, etc.

Boiler House, 1 story wooden building 40 x 33 ft.

Containing

1 Steam Boiler
1 Steam Pump
1 Steam Tube Cleaner
1 Round Tank, about 5 x 6 ft.

Coal Shed, 61 x 30 ft. Store, 61 x 16 ft. (Wooden Building)

Store Containing

4 Hatchets
10 Coal Shovels
4 Long Handled Shovels
1 Sledge
2 Pick Axes
4 Prongs
1 Bung Bore
2 Deck Brooms
1 White Head Light
1 Leather Hose
10 Wheel Barrows

Slip for Whales and Belongings, 110 x 60 ft.

1 Heaving Up Winch
1 Flencing Up Winch
1 Canting Winch
1 Heaving Up Tackle
2 Heaving Up Chains
6 Snatch Blocks

Slip for Steamer and Belongings

2 Heavy Heaving Up Tackles
2 Small Blocks
11 Big Shackles

Stable, Wooden Building 30 x 22 ft.

Poultry House, Wooden Building 14 x 19 ft.

Powder Magazine, Brick Building

Store, Wooden Building 12 x 14 ft.

Pig House, Wooden Building 11 x 24 ft.

Cellar, with Wooden Store on Top 16 x 20 ft.

Source: See appendix V.